Guide to Design Criteria for Bolted and Riveted Joints

Guide to Design Criteria for Bolted and Riveted Joints

John W. Fisher
John H. A. Struik

Fritz Engineering Laboratory
Lehigh University
Bethlehem, Pennsylvania

A WILEY-INTERSCIENCE PUBLICATION

JOHN WILEY & SONS
New York • Chichester • Brisbane • Toronto • Singapore

To our Families

Copyright © 1974, by John Wiley & Sons, Inc.

All rights reserved.

Reproduction or translation of any part of this work beyond that permitted by Sections 107 or 108 of the 1976 United States Copyright Act without the permission of the copyright owner is unlawful. Requests for permission or further information should be addressed to the Permissions Department, John Wiley & Sons, Inc.

Library of Congress Cataloging in Publication Data:

Fisher, John W   1931–
   Guide to design criteria for bolted and riveted joints.

   "A Wiley-Interscience publication."
   Includes bibliographical references.
   1. Bolted joints.   2. Riveted joints. I. Struik, John H. A., 1942–   joint author. II. Title.

TA492.B63F56   624′.182   73-17158
ISBN 0-471-26140-8

Printed in the United States of America

10 9 8 7 6 5

Foreword

From the earliest experiments which related to the strength of riveted connections by William Fairbairn 135 years ago, countless tests have been performed by a host of investigators seeking a better understanding of the behavior of riveted and bolted structural joints.

In his bibliography published in 1945 de Jonge lists some 1300 articles dealing with the subject. Since that time, either in *Bibliography on Bolted Joints*, sponsored by the Research Council on Riveted and Bolted Structural Joints, and published by the American Society of Civil Engineers in 1967 as *ASCE Manual on Engineering Practice,* No. 48, or as supplements appearing in the *Journal of the ASCE Structural Division*, more than 800 additional papers have been listed. The preponderance of this latter group are the outgrowth of a vast amount of research sponsored by Research Council on Riveted and Bolted Structural Joints since its formation in 1947.

The most significant single accomplishment resulting from this sponsored research was the development of the high-strength bolt. Provisions for their use are contained in the *RCRBSJ Specifications for Structural Joints Using ASTM A325 or A490 Bolts*, the Council's only publication to date. The promulgation of rules for the use of other types of mechanical fasteners was recognized as a prerogative of other specification-writing bodies, antedating the formation of the Council.

However, because of the accumulation of so much new information concerning the behavior of joints assembled with these latter fasteners, largely the result of the Council's research program, it was felt that a succinct digest of this material, within a single publication, would be a valuable aid to specification-writing bodies as well as to those engaged in the design and investigation of such connections. It is for this reason that this book has been prepared and approved by the Council for publication.

> T. R. HIGGINS
> Chairman, Subcommittee on Specifications
> Research Council on Riveted and Bolted
> Structural Joints

Preface

This book provides a state-of-the-art summary of the experimental and theoretical studies undertaken to provide an understanding of the behavior and strength of riveted and bolted structural joints. Design criteria have been developed on the basis of this information and should be beneficial to designers, teachers, students, and specification-writing bodies.

The book is intended to provide a comprehensive source of information on bolted and riveted structural joints as well as an explanation of their behavior under various load conditions. Design recommendations are provided for both allowable stress design and load factor design. In both cases, major consideration is given to the fundamental behavior of the joint and its ultimate capacity.

The work on this manuscript was carried out at Fritz Engineering Laboratory, Lehigh University, Bethlehem, Pa. The Research Council on Riveted and Bolted Structural Joints sponsored the project from its inception in 1969.

The work has been guided by the Councils Committee on Specifications under the chairmanship of Dr. Theodore R. Higgins. Other members of the committee include: R. S. Belford, E. Chesson, Jr., M. F. Godfrey, F. E. Graves, R. M. Harris, H. A. Krentz, F. R. Ling, W. H. Munse, W. Pressler, E. J. Ruble, J. L. Rumpf, T. W. Spilman, F. Stahl, and W. M. Thatcher. The authors are grateful for the advice and guidance provided by the committee. Many helpful suggestions were made during the preparation of the manuscript. Sincere appreciation is also due the Research Council on Riveted and Bolted Structural Joints and Lehigh University for supporting this work.

A book of this magnitude would not have been possible without the assistance of the many organizations who have sponsored research on riveted and bolted structural joints at Fritz Engineering Laboratory. Much of the research on the behavior of riveted and bolted structural joints that was conducted at Fritz Engineering Laboratory provided background for this study and was drawn on extensively. Those sponsoring this work include the American Institute of Steel Construction, the Pennsylvania Department of Transportation, the Research Council on Riveted and Bolted Structural Joints, the United States Department of Transportation-Federal Highway Administration, and the Louisiana Department of Transportation.

The authors are particularly grateful for the advice provided by Dr. Theodore R. Higgins and Dr. Geoffrey L. Kulak. Many helpful suggestions were provided that greatly improved the manuscript and design recommendations.

The manuscript was typed by Mrs. Charlotte Yost, and her assistance with the many phases of the preparation of the manuscript is appreciated. Acknowledgement is also due Mary Ann Yost for her assistance with the preparation of the various indexes provided in this book and other resource material. Many organizations have given permission to reproduce graphs, tables, and photographs. This permission is appreciated and credit is given at the appropriate place.

JOHN W. FISHER
JOHN H. A. STRUIK

Bethlehem, Pennsylvania
July 1973

Contents

1. **Introduction**   1

    1.1 Purpose and Scope, 1
    1.2 Historical Notes, 1
    1.3 Types and Mechanical Properties of Structural Fasteners, 3

2. **General Provisions**   9

    2.1 Structural Steels, 9
    2.2 Types of Connections, 11
    2.3 Loads, 16
    2.4 Factor of Safety—Load Factor Design, 17
    2.5 Bolted and Riveted Shear Splices, 18
    2.6 Fatigue, 20
    2.7 Fracture, 23
    2.8 Allowable Working Stresses, 26

3. **Rivets**   29

    3.1 Rivet Types, 29
    3.2 Installation of Rivets, 29
    3.3 Behavior of Individual Fasteners, 30

        3.3.1 *Rivets Subjected to Tension,* 31
        3.3.2 *Rivets Subjected to Shear,* 32
        3.3.3 *Rivets Subjected to Combined Tension and Shear,* 33

    3.4 Basis for Design Recommendations, 34

        3.4.1 *Rivets Subjected to Tension,* 35
        3.4.2 *Rivets Subjected to Shear,* 35
        3.4.3 *Rivets Subjected to Combined Tension and Shear,* 36

4. Bolts 37

    4.1 Bolt Types, 37
    4.2 Behavior of Individual Fasteners, 41

        4.2.1 Bolts Subjected to Tension, 41
        4.2.2 Bolts Subjected to Shear, 46
        4.2.3 Bolts Subjected to Combined Tension and Shear, 53

    4.3 Installation of High-Strength Bolts, 54
    4.4 Relaxation, 61
    4.5 Reuse of High-Strength Bolts, 62
    4.6 Galvanized Bolts and Nuts, 63
    4.7 Use of Washers, 65
    4.8 Corrosion and Embrittlement, 66
    4.9 Effect of Nut Strength, 67
    4.10 Basis for Design Recommendations, 67

        4.10.1 Bolts Subjected to Tension, 68
        4.10.2 Bolts Subjected to Shear, 68
        4.10.3 Bolts Subjected to Combined Tension and Shear, 69

5. Symmetric Butt Splices 71

    5.1 Joint Behavior Before Slip, 71

        5.1.1 Introduction, 71
        5.1.2 Basic Slip Resistance, 71
        5.1.3 Evaluation of Slip Characteristics, 72
        5.1.4 Effect of Joint Geometry and Number of Faying Surfaces, 74
        5.1.5 Joint Stiffness, 75
        5.1.6 Effect of Type of Steel, Surface Preparation, and Treatment on the Slip Coefficient, 76
        5.1.7 Effect of Variation in Bolt Clamping Force, 79
        5.1.8 Effect of Grip Length, 83

    5.2 Joint Behavior After Major Slip, 84

        5.2.1 Introduction, 84
        5.2.2 Behavior of Joints, 84
        5.2.3 Joint Stiffness, 88
        5.2.4 Surface Preparation and Treatment, 89

Contents

        5.2.5   *Load Partition and Ultimate Strength*, 90
        5.2.6   *Effect of Joint Geometry*, 94
        5.2.7   *Type of Fastener*, 107
        5.2.8   *Effect of Grip Length*, 107
        5.2.9   *Bearing Stresses and End Distance*, 108

  5.3 Joint Behavior Under Repeated Loading, 112

        5.3.1   *Basic Failure Modes*, 112
        5.3.2   *Fatigue Strength of Bolted Butt Joints*, 116

  5.4 Design Recommendations, 122

        5.4.1   *Introduction*, 122
        5.4.2   *Design Recommendations—Fasteners*, 123
        5.4.3   *Design Recommendations—Connected Material*, 133

6. Truss-Type Connections   143

  6.1 Introduction, 143
  6.2 Behavior of Truss-Type Connections, 145

        6.2.1   *Static Loading*, 145
        6.2.2   *Repeated Loading*, 150

  6.3 Design Recommendations, 151

7. Shingle Joints   154

  7.1 Introduction, 154
  7.2 Behavior of Shingle Joints, 154
  7.3 Joint Stiffness, 158
  7.4 Load Partition and Ultimate Strength, 158
  7.5 Effect of Joint Geometry, 159

        7.5.1   *Effect of Variation in A_n/A_s Ratio and Joint Length*, 159
        7.5.2   *Number of Fasteners per Region*, 160
        7.5.3   *Number of Regions*, 162

  7.6 Design Recommendations, 163

        7.6.1   *Approximate Method of Analysis*, 163
        7.6.2   *Connected Material*, 167
        7.6.3   *Fasteners*, 167

8. Lap Joints   169

    8.1 Introduction, 169
    8.2 Behavior of Lap Joints, 169
    8.3 Design Recommendations, 173

        8.3.1 Static Loading Conditions, 173
        8.3.2 Repeated Type Loading, 173

9. Oversize and Slotted Holes   174

    9.1 Introduction, 174
    9.2 Effect of Hole Size on Bolt Tension and Installation, 174
    9.3 Joint Behavior, 179

        9.3.1 Slip Resistance, 179
        9.3.2 Ultimate Strength, 181

    9.4 Design Recommendations, 181

10. Filler Plates between Surfaces   184

    10.1 Introduction, 184
    10.2 Types of Filler Plates and Load Transfer, 184
    10.3 Design Recommendations, 188

11. Alignment of Holes   190

    11.1 Introduction, 190
    11.2 Behavior of Joints with Misaligned Holes, 190
    11.3 Design Recommendations, 192

12. Surface Coatings   194

    12.1 Introduction, 194
    12.2 Effect of Type of Coating on Short-Duration Slip Resistance, 195

        12.2.1 Hot-Dip Galvanizing, 195
        12.2.2 Metallizing, 196
        12.2.3 Zinc-Rich Paints, 201

    12.3 Joint Behavior Under Sustained Loading, 205

Contents   xiii

    12.4   Joint Behavior Under Repeated Loading, 206
    12.5   Design Recommendations, 208

13. Eccentrically Loaded Joints   211

    13.1   Introduction, 211
    13.2   Behavior of a Fastener Group Under Eccentric Loading, 211
    13.3   Analysis of Eccentrically Loaded Fastener Groups, 214

        13.3.1   *Slip-Resistant Joints,* 215
        13.3.2   *Ultimate Strength Analysis,* 218

    13.4   Comparison of Analytical and Experimental Results, 219
    13.5   Design Recommendations, 220

        13.5.1   *Connected Material,* 220
        13.5.2   *Fasteners,* 221

14. Combination Joints   231

    14.1   Introduction, 231
    14.2   Behavior of Combination Joints Which Share Load on a Common Shear Plane, 233

        14.2.1   *High-Strength Bolts Combined with Welds,* 234
        14.2.2   *High-Strength Bolts Combined with Rivets,* 238

    14.3   Design Recommendations, 239

        14.3.1   *Static Loading Conditions,* 239
        14.3.2   *Repeated Loading Conditions,* 239

15. Gusset Plates   241

    15.1   Introduction, 241
    15.2   Method of Analysis and Experimental Work on Gusset Plates, 242
    15.3   Design Recommendations, 246

16. Beam and Girder Splices   250

    16.1   Introduction, 250
    16.2   Types and Behavior of Beam-Girder Splices, 250

    16.2.1 *Flange Splices,* 252
    16.2.2 *Web Splices,* 255

  16.3 Design Recommendations, 255

    16.3.1 *Flange Splices,* 255
    16.3.2 *Web Splices,* 256

17. Tension-type Connections   257

  17.1 Introduction, 257
  17.2 Single Fasteners in Tension, 257
  17.3 Bolt Groups Loaded in Tension—Prying Action, 260
  17.4 Repeated Loading of Tension-Type Connections, 266
  17.5 Analysis of Prying Action, 268
  17.6 Design Recommendations, 275

    17.6.1 *Static Loading,* 275
    17.6.2 *Repeated Loading,* 278

18. Beam-to-Column Connections   281

  18.1 Introduction, 281
  18.2 Classification of Beam-to-Column Connections, 282
  18.3 Behavior of Beam-to-Column Connections, 284

    18.3.1 *Flexible Beam-to-Column Connections,* 285
    18.3.2 *Semi-rigid Connections,* 288
    18.3.3 *Rigid Connections,* 289

  18.4 Stiffener Requirements for Bolted Beam-to-Column Connections, 294
  18.5 Design Recommendations, 299

Author Index   303

Subject Index   311

Guide to Design Criteria for Bolted and Riveted Joints

Chapter One
Introduction

1.1 PURPOSE AND SCOPE

The purpose of this book is to provide background information and criteria that can be used as a guide to the improvement of existing design procedures and specifications for bolted and riveted joints. To achieve this goal, extensive research work performed in the United States, Canada, Australia, Germany, the Netherlands, England, Norway, and Japan was reviewed.

Among the criteria considered as a basis for design was an evaluation of the load deformation characteristics of the component parts of the joint. Much emphasis was placed on the behavior of structural joints connected by ASTM-A325 or A490 high-strength bolts. The joint materials considered ranged from structural carbon steel with a specified yield stress between 33 and 36 ksi to quenched and tempered alloy steel with a yield stress ranging from 90 to 100 ksi.

The different types of fasteners, connections, loading conditions, and design procedures are discussed briefly in the first two chapters. Chapters 3 and 4 deal with the behavior of individual fasteners under various loading conditions. Chapter 5 describes the behavior, analysis, and design of symmetric butt splices. Special types of joints such as truss-type connections, shingle joints, beam girder splices, and beam-to-column connections are discussed in subsequent chapters.

1.2 HISTORICAL NOTES

Rivets were the principal fasteners in the early days of iron and steel, but occasionally bolts of mild steel were used in structures.[1.6, 1.8] It had long been known that hot driven rivets generally produced clamping forces. However, the axial force was not controlled and varied substantially. Therefore, it could not be evaluated for design.

Batho and Bateman were the first to suggest that high-strength bolts could be used to assemble steel structures.[1.1] In 1934 they reported to the Steel Structures Committee of Scientific and Industrial Research of Great Britain that bolts could be tightened enough to prevent slip in structural joints. It was concluded that bolts with a minimum yield strength of 54 ksi could be tightened sufficiently to give an adequate margin of safety against slippage of the connected parts.

Based on tests performed at the University of Illinois, Wilson reported[1.2] in 1938:

> The fatigue strength of high-strength bolts appreciably smaller than the holes in the plates was as great as that of well driven rivets if the nuts were screwed up to give a high tension in the bolt.

Little more was done about high-strength bolting until 1947 when the Research Council on Riveted and Bolted Structural Joints (RCRBSJ) was formed. The purpose of the Council was as follows:

> To carry on investigations as many as eemed rnecessary tc determine the suitability of various types of joints used in structural frames.

The Council sponsored studies on high-strength bolts and rivets and their use in structural connections. The realization that bolts could be extremely useful in the maintenance of bridges helped support developmental work at this early stage. The use of high-strength steel bolts as permanent fasteners has become general since the formation of the RCRBSJ. Prior to that time heat-treated carbon bolts were only used for fitting-up purposes and for carrying the loads during erection. The bolts were tightened to pull the plies of joint material together. No attempt was made to attain a precise amount of clamping force.

The American Society for Testing and Materials (ASTM) in conjunction with the RCRBSJ prepared a tentative specification for the materials for high-strength bolts, a specification which was first approved in 1949.[1.3] Using the results of research, the RCRBSJ prepared and issued its first specification for structural joints using high-strength bolts in January 1951.[1.4] This specification permitted the rivet to be replaced by a bolt on a one-to-one basis.

In the early 1950s, the installation procedures, the slip resistance of joints having different surface treatments, and the behavior of joints under repeated loadings were studied.[1.6] Outside of the United States high-strength bolts also attracted much attention. Sufficient experience was gained in the laboratory and in bridge construction to enable the German Committee for Structural Steelwork (GCSS) to issue a preliminary code of practice (1956).[1.7] In Great Britain, the general practice was similar to practice and specifications in the United States. The British Standards Institution issued a British Standard (BS) 3139 dealing with bolt material in 1959. In 1960, BS 3294 was issued to establish the design procedure and field practice.

1.3 Types and Mechanical Properties of Structural Fasteners

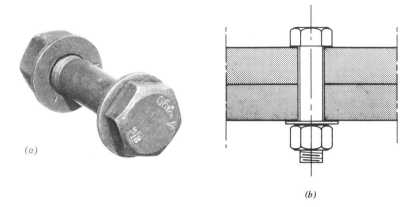

Fig. 1.1. Heavy hex bolts. (*a*) High-strength bolt (Courtesy of Bethlehem Steel Corp.); (*b*) installed bolt.

Research developments led to several editions of the RCRBSJ specifications. Allowable stresses were increased, tightening procedures modified, and new developments such as the use of A490 alloy steel bolts, galvanized joints and bolts, and slotted holes were incorporated.[1.4]

1.3 TYPES AND MECHANICAL PROPERTIES OF STRUCTURAL FASTENERS

The mechanical fasteners used in structural connections can be classified as either rivets or bolts. Both serve the same purpose, but there are significant differences in appearance. Standards for both types of fasteners are given in Ref. 1.5.

The most commonly used types of structural bolts are (1) the ASTM A307 grade A low carbon steel bolt, (2) the ASTM A325 high-strength steel bolt, and (3) the ASTM A490 alloy steel bolt.[1.3, 1.9, 1.10]

ASTM designation	Bolt diameter (in.)	Tensile strength[a] (ksi)
A307-68	All	60 minimum
A325-70a	$\frac{1}{2}$–1	120 minimum
	$1\frac{1}{8}$–$1\frac{1}{2}$	105 minimum
A490-70a	$\frac{1}{2}$–$1\frac{1}{2}$	150 minimum–170 maximum

Fig. 1.2. Tensile strength requirements of structural bolts. [a] Computed on the stress area.

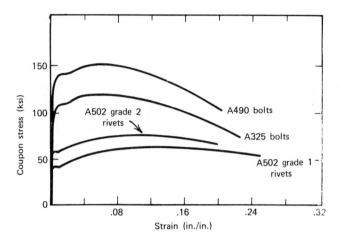

Fig. 1.3. Coupon stress-strain relationships for different fastener materials.

The ASTM A307 low carbon steel fastener is primarily used in light structures, subjected to static loads. The high-strength A325 and A490 bolts are heavy hex structural bolts, used with heavy hex nuts (see Fig. 1.1).

A307 bolts are made of low carbon steel with mechanical properties as designated by ASTM A307. A325 bolts are made by heat-treating, quenching, and tempering medium carbon steel. Two different strength levels are specified depending on the size of the bolts (see Fig. 1.2).[1.3] The quenched and tempered alloy steel bolt, designated as A490 bolt, has higher mechanical properties as compared to the A325 high-strength bolt. It was

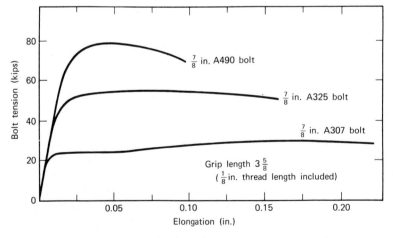

Fig. 1.4. Comparison of bolt types (direct tension).

1.3 Types and Mechanical Properties of Structural Fasteners

especially developed for use with high-strength steel members. The A490 specification calls for the heavy head and the short thread length of the A325 specification together with chemical and physical properties nearly identical to the A354 grade BD bolt.[1.11] For the development of the A490 bolt many calibration tests were performed on A354 grade BD fasteners manufactured to conform to the A490 specification requirements. The mechanical properties of the different bolt types for structural joints are summarized in Figs. 1.2 and 1.3. Unlike rivets, the strength of bolts is specified in terms of a tensile test of the threaded fastener. The load-elongation characteristics of a bolt are more significant than the stress-strain diagram of the parent metal because performance is controlled by the threads. Also, the stress varies along the bolt as a result of the gradual introduction of force from the nut and the change in section from the threaded to the unthreaded portion. The weakest section of any bolt in tension is the threaded portion. The tensile strength of the bolt is usually determined from the "stress area" defined as:

$$\text{stress area} = 0.785 \left(D - \frac{0.9743}{n} \right)^2$$

where D = nominal bolt diameter
n = number of threads per inch

Figure 1.4 shows typical load-elongation curves for three different bolts of the same diameter. The tensile strength of each of the bolts was near minimum specified.

In addition to regular structural bolts, threaded parts have many other structural applications, for example, anchor bolts or tension rods. Anchor bolts are used in column base plates to prevent the uplift of the base plate due to column moments. Threaded parts in tension rods are frequently used to transmit tensile loads from one element to another. In all of these applications the threaded parts are primarily subjected to tensile loads and the ultimate tensile load of these connections is determined on the basis of the stress area.

The nut is an important part of the bolt assembly. Nut dimensions and strengths are specified so that the strength of the bolt is developed.[1.5]

Bolts are generally used in holes $\frac{1}{16}$ in. (2 mm) larger than the nominal bolt diameter. When A307 or other mild steel bolts are used, the connection is commonly in bearing and the nuts are tightened sufficiently to prevent play in the connected members. The clamping force is not very great and should not be considered in design. High-strength bolts (A325 and A490) can produce high and consistent preloads and are required to be tightened to a tension equal to or greater than 70% of the minimum tensile

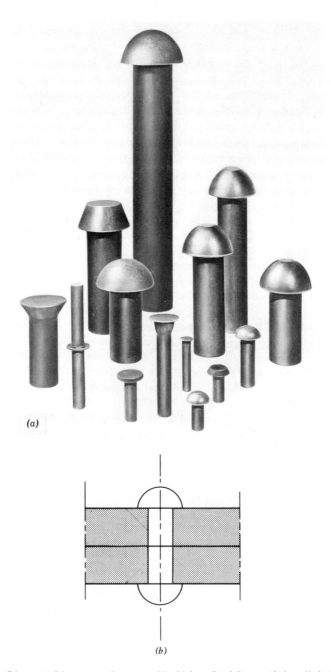

Fig. 1.5. Rivets. (a) Rivet types (Courtesy of Bethlehem Steel Corp.); (b) installed rivet.

	Grade 1		Grade 2	
	Min.	Max.	Min.	Max.
Rockwell B	55	72	76	85
Brinell, 500-kg load, 10-mm ball	103	126	137	163

Fig. 1.6. Hardness requirements for A502 rivet steel.

strength. Such tightening requires the use of hand torque-wrenches or powered impact wrenches. Two methods of controlling bolt tension are used. A detailed description of both tightening procedures is given in Chapter 4.

Rivets are made from bar stock by either hot- or cold-forming the manufactured head. The head is usually of the high button type although flattened and countersunk rivets are made for applications with limited clearance. Different rivet types are shown in Fig. 1.5.

Structural rivet steels are mainly of two types: (1) ASTM A502 grade 1, low carbon rivet steel, and (2) ASTM A502 grade 2, high-strength structural steel rivets.[1.12] Grade 1 and 2 rivets correspond to those formerly made from steel conforming to ASTM A141 and A195, respectively. The mechanical hardness requirements for A502 rivet steel are listed in Fig. 1.6. The stress-strain relationships for typical, undriven A502 rivets are given in Fig. 1.3. For comparative purposes this figure also shows the stress strain curves obtained from 0.505-in. diameter specimens turned from full size A325 and A490 bolts.

References

1.1 C. Batho and E. H. Bateman, *Investigations on Bolts and Bolted Joints*, second report of the Steel Structures Research Committee, London 1934.

1.2 W. M. Wilson and F. P. Thomas, *Fatigue Tests on Riveted Joints*, Bulletin 302, Engineering Experiment Station, University of Illinois, Urbana, 1938.

1.3 American Society for Testing and Materials, *High-Strength Bolts for Structural Steel Joints, Including Suitable Nuts and Plain Hardened Washers*, ASTM Designation A325–70a (originally issued 1949), Philadelphia, 1970.

1.4 Research Council on Riveted and Bolted Structural Joints of the Engineering Foundation, *Specifications for Assembly of Structural Joints Using High-Strength Bolts*, originally issued 1951, latest edition, 1972; *Specifications for Structural Joints Using ASTM A325 or A490 Bolts*.

1.5 Industrial Fasteners Institute, *Fastener Standards*, 5th ed., Industrial Fasteners Institute, Cleveland, Ohio, 1970.

1.6 ASCE-Manual 48, *Bibliography on Bolted and Riveted Joints*, Headquarters of the Society, New York, 1967.

1.7 Deutscher Stahlbau-Verband, *Preliminary Directives for the Calculation, Design and Assembly of Non-Slip Bolted Connections*, Stahlbau Verlag, Cologne, 1956.

1.8 A. E. R. de Jonge, *Bibliography on Riveted Joints*, American Society of Mechanical Engineers, New York, 1945.

1.9 American Society for Testing and Materials, *Quenched and Tempered Alloy Steel Bolts for Structural Joints*, ASTM Designation A490–70, Philadelphia, 1970.

1.10 American Society for Testing and Materials, *Low Carbon Steel Externally and Internally Threaded Standard Fasteners*, ASTM Designation A307–68, Philadelphia, 1968.

1.11 American Society for Testing and Materials, *Quenched and Tempered Alloy Steel Bolts and Studs with Suitable Nuts*, ASTM Designation A354–66, Phaladelphia, 1966.

1.12 American Society for Testing and Materials, *Steel Structural Rivets*, ASTM Designation A502–65, Philadelphia, 1966.

Chapter Two

General Provisions

2.1 STRUCTURAL STEELS

Knowledge of the material properties is a major requirement for the analysis of any structural system. The strength and ductility of a material are two characteristics needed by the designer. These material properties are often described adequately by the stress-strain relationship for the material. Figure 2.1 shows the stress-strain relationship that is characteristic of many steels for structural applications. The figure shows the four typical ranges of behavior: the elastic range, the plastic range (during which the material flows at a constant stress), the strain-hardening range, and the range during which necking occurs terminating in fracture. Generally the initial elastic and yield segments are the most important portions. The following points can be noted in Fig. 2.1:

1. Over an initial range of strain, stress and strain are proportional. The slope of the linear relationship is Young's modulus E.

2. After the initiation of yield there is a flat plateau. The extent of the yield zone (or "plastic range") can be considerable.

3. At the end of the plateau, strain-hardening begins with a subsequent increase in strength.

Structural steel can undergo sizable permanent (plastic) deformations before fracture. In contrast to a brittle material, it will generally show signs of distress through permanent but noncatastrophic plastic deformation. The energy absorbed during the process of stretching is proportional to the area under the stress-strain curve. The ductility is essential in various ways for the proper functioning of steel structures, particularly that of connections.

Structural steels can be classified as follows:

1. Structural carbon steel with a specified yield stress between 33 and 36 ksi. Typical examples are A36 and Fe37 steels.

2. High-strength steel with a specified yield stress between 42 and 50 ksi. A typical example in this category is A440 steel.

3. High-strength low alloy steels with a specified yield stress ranging from 40 to 65 ksi. This category comprises steels such as A242. A441, A572, A588, and Fe52.

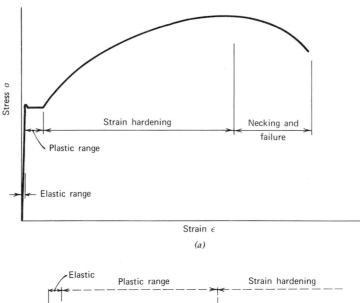

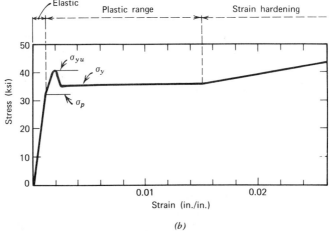

Fig. 2.1. Stress-strain curve. (*a*) Stress-strain curve for structural carbon steel; (*b*) initial portion of stress-strain curve.

4. Quenched and tempered carbon steel with a specified yield stress between 50 and 60 ksi. A537 steel is a typical example.

5. Quenched and tempered alloy steel with a specified yield stress ranging from 90 to 100 ksi. Materials in this category are covered by ASTM A514 and A517.

Typical stress-strain curves for these steels are given in Fig. 2.2. The curves shown are for steels having specified minimum tensile properties. The corresponding properties of these steels are listed in Fig. 2.3.

2.2 Types of Connections

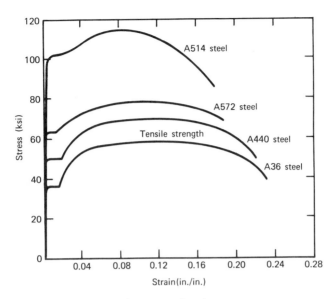

Fig. 2.2. Typical stress-strain curves for structural steels.

2.2 TYPES OF CONNECTIONS

Mechanically fastened joints are conveniently classified according to the type of forces to which the fasteners are subjected. These classes are (1) shear, (2) tension, and (3) combined tension and shear. Under category 1 the fasteners are loaded either in axial or eccentric shear. If the line of action of the applied load passes through the centroid of the fasteners group, then the fasteners are loaded in axial shear. In eccentric shear the

Steel Type	Minimum Yield Stress (ksi)	Tensile Strength (ksi)	Minimum Elongation in 8 in.[b] (%)
A36-70a	36	58–80	20
A440-70a	42–50[a]	63–70[a]	18
A242-70a	42–50[a]	63–70[a]	18
A441-70a	42–50[a]	60–70[a]	18
A572-70a	42–65[a]	60–80[a]	15–20[a]
A588-70a	42–50[a]	63–70[a]	18
A537-70	50–60	70–100	18
A514-70	90–100[a]	105–135[a]	17–18[a]

Fig. 2.3. Minimum specified properties for structural steels. [a] Depending on thickness. [b] 2 in. for A514 steel.

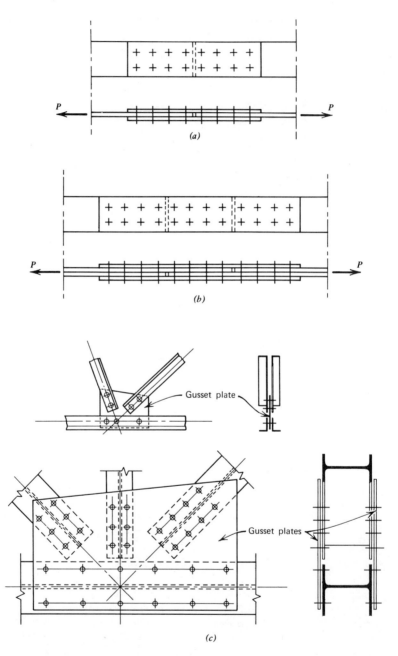

Fig. 2.4. Typical riveted and bolted connections. (*a*) Symmetric butt splice; (*b*) shingle splice; (*c*) single plane construction (top); double plane construction (bottom); (*d*) lap splice; (*e*) bracket connection; (*f*) girder web splice; (*g*) hanger connection.

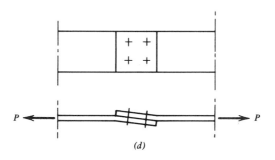

(d)

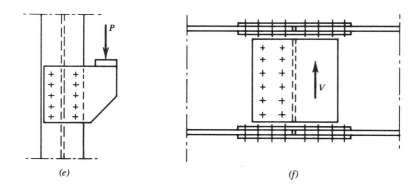

(e)   *(f)*

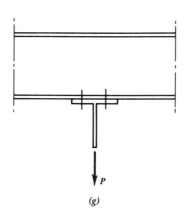

(g)

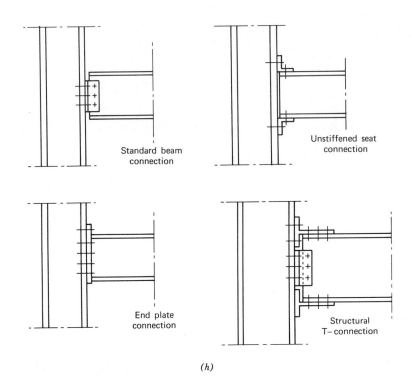

(h)

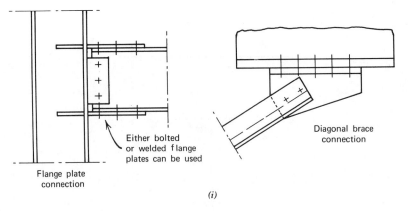

(i)

Fig. 2.4 cont.

2.2 Types of Connections

shear force does not pass through the centroid of the fastener group. This results in an additional torsional moment on the fastener group that increases the fastener shear stresses. This loading condition is referred to as eccentric shear.

The simplest type of structural connection subjecting fasteners to axial shear are flat plate-type splices. Typical examples are shown in Fig. 2.4a, b, and c. The butt splice is the most commonly used because symmetry of the shear planes prevents bending of the plate material. The load is applied through the centroid of the fastener group. The fasteners act in double shear, since two shearing planes cross the fastener.

Instead of a symmetric butt splice, the shingle splice (Fig. 2.4b) may be used when the main member consists of several plies of material. A more gradual transfer of load in the plate occurs with this staggered splice than if all main plates are terminated at the same location.

Other examples of joints in which the fasteners are subjected to axial shear are gusset plate connections. Depending on the joint geometry, the fasteners are subjected to either double or single shear as illustrated in Fig. 2.4c. Generally bending is prevented even though the fasteners are in single shear, because of symmetry of the two shearing planes.

In the lap plate splice shown in Fig. 2.4d the fasteners act in single shear. The eccentricity of the loads pulling on the connected members causes bending as the loads tend to align axially. Because of these induced bending stresses, this type of connection is only used for minor connections.

Often situations arise in which the line of the force acting on a connection does not pass through the centroid of the fastener group. This implies that the fastener groups are subjected to eccentric shear forces. Typical examples in this category are bracket connections and web splices of plate girders as shown in Figs. 2.4e and f.

A hanger type connection (Fig. 2.4g) is one of the few examples where mechanical fasteners are used in direct tension. More often fasteners are subjected to the combined action of tension and shear. This is common in building frames and bridge deck systems if the connections are required to transmit moments to ensure continuous structural action. The amount of continuity depends on the ability of the connection to resist moments. Moment connections may produce conditions where the upper fasteners are being loaded in shear by the vertical reaction and loaded in tension by the end moment. Some examples of frame connections are given in Fig. 2.4h. Another type of connection in which the fasteners are subjected to combined tension and shear is the diagonal brace shown in Fig. 2.4i.

The behavior, analysis, and design of the four major categories of connections—fasteners loaded in axial shear, eccentric shear, tension, or combined tension and shear—are discussed in the following chapters.

2.3 LOADS

The loads and forces acting on a structure may be divided into two broad categories: (1) dead loads and (2) live loads or forces. Dead loads are static, gravitational forces. For a building this usually includes the weight of the permanent equipment and the weight of the fixed components of the building such as floors, beams, girders, and such. In a bridge it includes the weight of the structural frame, wearing surfaces, lighting fixtures, and such.

As contrasted to the dead loads on a structure, the magnitude of live loads is generally variable with time. Also, most dead loads are static loads. Live loads are often dynamic loads. In many situations the dynamic nature of the forces has only minor influence on the stress distribution and these loads can be treated as statically applied loads. Live loads can be subdivided into vertical and lateral live loads. The loads on a building due to its occupancy as well as snow loads on roof surfaces are regarded as vertical live loads. These load provisions are usually specified in local building codes. Live loads on bridges depend on usage and are specified in the relevant codes such as the AREA[2.1] code for railway bridges and the AASHO[2.2] specifications which are applicable to highway bridges.

If live loads are dynamic in nature such as moving vehicles on a bridge or a hoisting machine in a building, it is necessary to account for their dynamic or impact effects. It is well known that the momentum of the load produces internal forces above the static values. In such situations the design load is equal to the sum of he dead load D, the live load L, and the impact load I. The total effect of live load and impact load is usually evaluated by multiplying the live load L by an impact factor p, where p is larger than 1.0. The fraction of p in excess of 1.0 accounts for the load increase because of the dynamic nature of the live load. The impact factor p depends on the type of member, its dimensions, and its loading condition. The factor p is usually prescribed in relevant codes.

Lateral live loads include earth or hydrostatic pressure on the structure and the effects of wind and earthquakes. It also includes the centrifugal forces caused by moving loads on curved bridges.

Wind is normally treated as a statically applied pressure, neglecting its dynamic nature. This is justified mainly on the basis of lack of significant periodicity in the fluctuating wind. However, experience has shown this procedure to be unacceptable in certain types of structures such as suspension bridges and other flexible structures where special consideration of dynamic wind effects is essential.

An earthquake is a ground motion caused by a sudden fracture and slidings along the fractured surface of the earth crust. Earthquakes are volcanic or tectonic in origin. The forces developed during an earthquake are

2.4 Factor of Safety—Load Factor Design

internal forces resulting from the tendency of the structure to resist motion. The structure should be capable of resisting these forces with a sufficient margin of safety against distress, that is, full or partial failure or excessive deformations. Some codes, such as the SEAOC[2.3] code, present practical minimum earthquake design procedures for typical structures. In special types of structures a more elaborate analysis of the dynamic response of the structure may be required.

Member forces can also result from temperature effects and support settlements. Consideration must also be given during the design to erection loads.

2.4 FACTOR OF SAFETY—LOAD FACTOR DESIGN

Failure of a structural connection occurs when the externally applied loads exceed the load-carrying capacity (ultimate load). The capacity of a connection can be based on strength or performance criteria. In the first case, loads in excess of the ultimate load lead to a complete or partial collapse of the connection. If performance is the controlling factor, the connection may lose its serviceability before its load capacity is reached because of excessive deformations, fatigue, or fracture. In this respect unrestricted plastic flow in a structural component is often regarded as determining the useful ultimate load of the member.

Structural members and connections are designed to have reserve beyond their ordinary service or working load. Allowance must be made for factors such as the variation in quality of materials and fabrication, possible overloads, secondary stresses due to errors introduced by design assumptions, and approximations in calculation procedures. In current design practice, a factor of safety is usually employed to provide for these uncertainties. In allowable stress design, the stress (or load) at failure is reduced by a factor of safety. This method does not account directly for the statistical nature of the design variables. The expected maxima of loading and the minima of strength not only are treated as representative parameters for design, but also are assumed to occur simultaneously. Neglecting the magnitude and frequency relationships for loads and strengths usually leads to conservative designs. It also results in different reliabilities for some safety factor.

A different approach to the problem of structural safety can be made by employing the concept of failure probabilities.[2.4, 2.13, 2.17-2.19] When the stress and strength distributions are known, structural safety may be determined from the probability that the stress due to the applied load will exeed the strength of a member as illustrated in Fig. 2.5. The shaded area indicates a finite probability of failure. As the overlap increases, the shaded area, and consequently the failure probability, increases proportionally. Hence changes in failure probability accompany changes in the stress-strength dis-

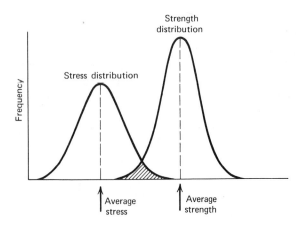

Fig. 2.5. Stress and strength distributions, including probability of failure.

tribution overlap. By employing the failure probability concept, a uniform reliability throughout the structure can be achieved.

The failure probability of a structural component is considered in a simplified way by the load factor design method.[2.22, 2.23] An expression for the maximum strength of a connection can be equated to the strength required to resist the various forces to which it will be subjected. The forces are increased by suitable factors intended to offset uncertainties in their magnitude and application. Thus

$$\Phi S = \alpha D + \gamma (L + I)$$

where S represents the average strength, D equals the dead load, and $L + I$ is the summatin of the live load and impact load on the connection. The factor Φ relates to uncertainties in the strength of the connection, whereas the factors α and γ relate to the chance of an increase in load. The factor Φ is evaluated from a strength distribution curve. The factors α and γ are determined from the distribution curves for dead load and the summation of live load and impact, respectively.

The design recommendations given in the following chapters have been developed considering both the factor of safety concept and the probabilistic approach used in load factor design.

2.5 BOLTED AND RIVETED SHEAR SPLICES

In Section 2.2, different types of connections were classified according to the type of forces to which the fasteners are subjected. If the fasteners in a joint are subjected to shear loads, a further classification based on connection performance is often made. This is illustrated by the behavior of the

2.5 Bolted and Riveted Shear Splices

symmetric butt joint shown in Fig. 2.6. The fasteners can be rivets or bolts with the clamping force provided by tightening the bolts or shrinkage of the rivets due to cooling. If the joint is subjected to an in-plane load through the centroid of the fasteners group, four characteristic loading stages exist as illustrated in Fig. 2.6. In the first stage, static friction prevents slip; in the second stage, the load exceeds the frictional resistance and the joint slips into bearing; in the third stage, the fasteners and plates deform elastically, and consequently the load-deformation relationship remains linear; in the fourth stage, yielding of plates, fasteners, or both occurs and results in plate fracture or complete shearing of the fasteners. Overlapping effects may make the distinctions between stages less clear-cut than depicted; however, in many tests these stages can be recognized clearly.

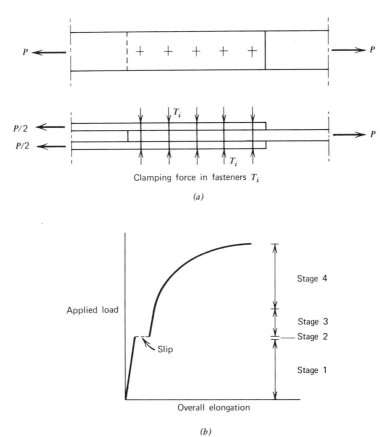

Fig. 2.6. Typical load elongation curve of symmetrical butt joint. (a) Symmetric butt joint; (b) load-elongation.

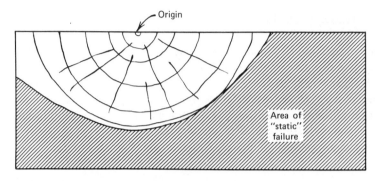

Fig. 2.7. Diagrammatic representation of a typical fatigue fracture surface.

In splices subjected to shearing loads, two methods of load transfer are possible: (1) by friction, and (2) by shear and bearing.

If the load on the connection is completely transferred by the frictional resistance on the contact surfaces, it is a slip resistant joint. Since slip does not occur, these connections are appropriate in situations where slip of the connection is not acceptable (i.e., repeated reversed stress conditions or situations where slip would result in undesirable misalignment of the structure). In slip resistant joints, the fasteners are not actually stressed in shear, and bearing is not a consideration.

If slip is not considered a critical factor, a load transfer by shear and bearing is acceptable. Joint slip may occur before the working load of the connection is reached depending on the available slip resistance. Slip brings the connected parts to bear against the sides of the fasteners, and the applied load is then transmitted partially by frictional resistance and partially by shear on the fasteners depending on joint geometry.

High-strength bolts are very suitable for use in slip resistant joints, since the magnitude of the axial bolt clamping force, which affects directly the frictional resistance of the connection, can be readily controlled. This is not true for rivets. Although a clamping force may be developed it is not reliable. Therefore, riveted joints are usually considered as bearing-type joints.

2.6 FATIGUE

Many structural members may be subjected to frequently repeated cyclical loads. Experience has shown that members and connections under such conditions may eventually fail from fatigue or stable crack growth even though the maximum applied stress is less than the yield stress. In general, fatigue failures occur when the nominal cyclic stress in the member is much lower than the elastic limit. These failures generally show little evidence of

2.6 Fatigue

deformation. Because of this lack of deformation, fatigue cracks are difficult to detect until substantial crack growth has occurred.

A fatigue fracture surface normally presents a characteistic appearance with three distinct and recognizable regions. The first region corresponds to slow stable crack growth. This has a visually smooth surface. The second region is rougher in texture as the distance and rate of growth from the nucleus of the fatigue crack increases. The third region is the final fracture, which may be either brittle or ductile depending on circumstances. Figure 2.7 shows schematically the different stages of a fatigue crack.

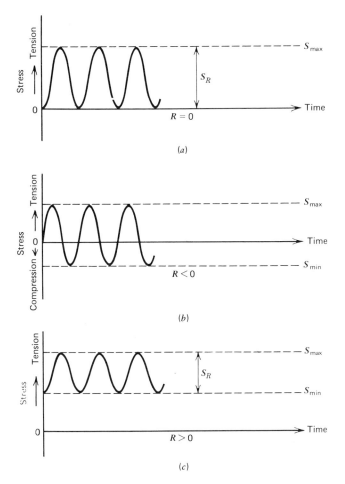

Fig. 2.8. Typical load cycles for fatigue testing. (*a*) Pulsating tension; (*b*) alternating; (*c*) fluctuating.

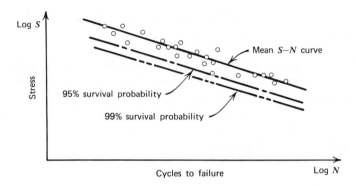

Fig. 2.9. *S-N* curve and corresponding survival probability curves.

For mechanically fastened joints, fatigue crack growth usually starts on the surface at a point of stress concentration such as a hole, a notch, a sharp fillet, a point of fretting, and such. Notches and other discontinuities cause stress rising effects immediately around the notch and decrease the fatigue strength. The elastic stress concentration factor for an infinitely wide plate with a circular hole and subjected to uniaxial uniform tension is equal to 3.0. Reducing the width of the plate as well as transmitting the load into the plate through a pin type loading at the hole increases the stress concentration factor significantly. Hence a change of shape results in a reduction in cross-sectional area and the type of load transfer, which are both significant factors that influence the magnitude of stress concentrations.

The only way to obtain a quantitative measure of the fatigue strength is to carry out fatigue tests under controlled conditions. The load cycle can be one of the following basic types: (*a*) pulsating, (*b*) reversing, and (*c*) fluctuating. Corresponding typical diagrams are shown in Fig. 2.8. In the past the loading cycle has often been characterized by the stress ratio R (algebraic ratio S_{min}/S_{max}) and the maximum stress; $R = 0$ corresponds to a pulsating load whereas a negative R represents a stress reversal type loading. A positive R corresponds to a fluctuating type loading. Recent work has indicated that the stress range is the dominant stress variable causing crack growth. [2.14, 2.15, 2.21]

For most structural joints it is necessary to actually test prototype specimens to evaluate the fatigue strength. The resulting test data provide a relationship between the applied stress and the number of cycles to failure. This relationship is referred to as an *S-N* curve and it forms a basis for describing the fatigue behavior of certain types of joints.

It has been shown in the literature that the *S-N* curve is well represented by a straight line when a logarithmic transformation of cycle life and

2.7 Fracture

maximum stress or stress range is made.[2.5-2.7] If sufficient data are available, a mean S-N curve can be determined as illustrated in Fig. 2.9. This line represents the 50% survival probability of all specimens. The tolerance limits of the S-N curve can be developed from the variation and survival probability. The desired level of survival probability can be used to develop design stresses for any number of applied stress cycles. Such a procedure is used in Chapter 5.4 to evaluate design recommendations for bolted joints subjected to repeated loadings.

In recent years the fracture mechanics of stable crack growth has confirmed the suitability of an exponential relationship between cycle life and applied stress range.[2.14, 2.15] The tool is expected to be of considerable help in evaluating the fatigue behavior of joints.

2.7 FRACTURE

As the temperature decreases, an increase is generally noted in the yield stress and tensile strength of structural steels. In contrast, the ductility usually decreases with a decreasing temperature. Furthermore, there is usually a temperature below which a specimen subjected to tensile stresses may fracture by cleavage. Little or no plastic deformation is observed in contrast to shear failure which is usually preceded by a considerable amount of plastic deformation. Both types of failure surfaces are shown in Fig. 2.10. Fractures that occur by cleavage are commonly referred to as brittle failures and are characterized by the propagation of cracks at very high velocities. There is little visible evidence of plastic flow and the fracture surface often appears to be granular except for thin portions along the edges.

Brittle fractures may be initiated at relative low nominal stress levels provided certain other conditions are present, such as (*a*) a flaw (a fatigue crack or a fabrication crack due to punched holes, etc.), (*b*) a tensile stress of sufficient intensity to cause a small deformation at the crack or notch tip, and (*c*) a low toughness steel that promotes cleavage fracture at the notch tip (a low service temperature will further aggravate this condition).

To understand brittle fracture one must look at the effects of stress concentrations accompanied by constraints that prevent plastic redistribution of stress. This is the condition that exists in the axially loaded notched bar shown in Fig. 2.11. The stress concentration effect of the notch or crack tip causes high longitudinal stresses at the apex of the notch and lower longitudinal stresses in the adjacent material. The lateral contraction in the width and thickness direction of the highly stressed material at the apex of the notch is restrained by the smaller lateral contraction of the lower stressed material. Consequently, tensile stresses are induced in the width and thickness directions (x and z) so that a severe triaxial state-of-

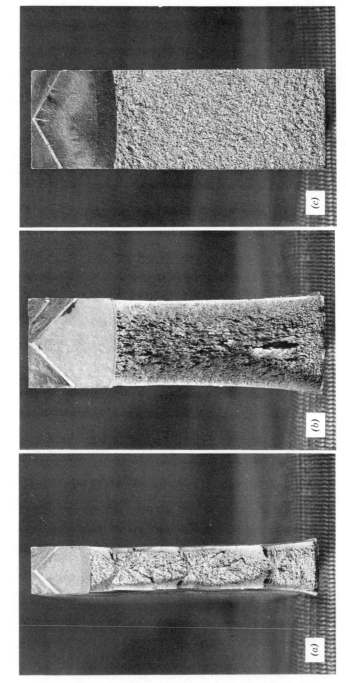

Fig. 2.10. Typical ductile and brittle fracture surfaces. (*a*) Ductile fracture surface with shear lip; (*b*) transition fracture surface; (*c*) brittle fracture surface with flat cleavage fracture.

2.7 Fracture

stress is present near the crack tip. Under these conditions a cleavage-or brittle-type failure may occur.

With decreasing temperatures, the transition from ductile behavior at the crack tip to cleavage behavior occurs within in a narrow temperature range. Usually the Charpy V-notch test is used to evaluate the suspectibility of a steel to brittle fracture. However, in this approach, important factors such as the flaw size and the stress concentration factors are not taken into account. These factors can be accounted for if a fracture analysis or fracture diagram is used.[2.8] The fracture diagram combines fracture mechanics, stress concentration factors, and flaw size with the transition temperature test approach. A detailed description of this concept is given in Refs. 2.8 through 2.10. Considerable work is in progress to assist with the development of fracture mechanics procedures that can be used to define fracture instability conditions. A correlation between the Charpy V-notch and K_{IC}, the plane-strain stress intensity factor at onset of unstable crack growth has been suggested.[2.20]

It is apparent that special attention must be directed to design and fabrication details of mechanically fastened connections so that brittle fractures will be avoided. A structural steel with a stable crack growth rate under service conditions should be selected.

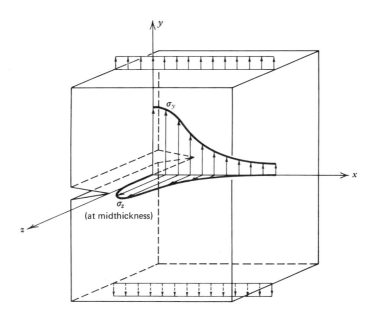

Fig. 2.11. State of stress at the root of a notch under uniaxial loading. *Note.* σ_y induces σ_z and σ_x stresses. The latter one is not shown in this figure.

One of the critical details in a bolted or riveted structure are the fasteners' holes. Punching the holes causes strain aging and work hardening of the material around the hole. Minute cracks radiating from the hole may form in the work hardened material, resulting in a notch in a region of high tensile stresses.[2.16]

To eliminate these potential points of crack initiation, punched holes should be reamed to remove the work-hardened material if brittle fracture is possible under service loads. Furthermore, geometrical discontinuities such as abrupt changes in cross-section should be avoided.

2.8 ALLOWABLE WORKING STRESSES

Allowable stress design is performed by specifying working loads and allowable stresses. Based on this concept, allowable tension and shear and bearing stresses for different types of fasteners are defined in the various specifications.[2.1, 2.2, 2.11, 2.12] Usually these stresses are applicable to the nominal bolt area, that is, the area corresponding to the nominal bolt diameter.

Although the bolts are not actually subjected to shear in slip resistant joints since the load is transmitted by frictional forces on the faying surfaces, it is convenient to specify an "allowable shear stress" for this category of joints. This permits the designer to use the same technique for all "shear type" splices. The allowable shear stress in a slip resistant joint must always be below, or at best equal to the allowable shear stress permitted on the basis of joint strength. The allowable shear stress on a fastener can be represented as

$$\tau_a = \beta \tau_{basic}$$

where τ_{basic} is the allowable shear stress based on joint strength and the minimum required factor of safety; β is a factor, less than or equal to one, taking into account the performance and behavior of the joint and can be written as

$$\beta = F(\beta_1, \beta_2, \beta_3)$$

where β_1 relates to the frictional resistance of the joint (surface treatment, slip coefficient, etc.) and the acceptable slip probability, β_2 describes the influence of the type of tightening procedure, and β_3 represents the influence of factors related to the specific joint geometry. Factors such as the influence of oversize and or slotted holes on the slip behavior of a joint are included as well as joint length.

This concept makes it possible to evaluate allowable stresses for various surface conditions taking into account specific joint performance requirements. Its application is discussed in greater detail in subsequent chapters.

References

2.1 American Railway Engineering Association, *Specifications for Steel Railway Bridges*, Chicago, 1969.

2.2 American Assoication of State Highway Officials, *Standard Specifications for Highway Bridges*, 11th ed., Washington, D.C., 1973.

2.3 Structural Engineers Association of California, *Recommended Lateral Force Requirements and Commentary*, Seismology Committee SEAOC, San Francisco.

2.4 E. B. Haugen, *Probabilistic Approaches to Design*, Wiley, New York, 1968.

2.5 W. Weibull, *Fatigue Testing and Analysis of Results*, Pergamon, New York, 1961, pp. 174–178, 192–201.

2.6 American Society for Testing and Materials, *A Guide for Fatigue Testing and Statistical Analysis of Fatigue Data*, ASTM Special Technical Publication 91-A, Philadelphia, 1963.

2.7 H. S. Reemsnyder, "Procurement and Analysis of Structural Fatigue Data," *Journal of the Structural Division, ASCE*, Vol. 95, ST7, July 1969.

2.8 W. S. Pellini, "Principles of Fracture Safe Design," *Welding Journal*, Vol. 50, No. 3 and 4, March–April 1971.

2.9 American Society for Testing and Materials, *Fracture Toughness Testing*, ASTM Special Technical Publication 381, Philadelphia, 1965.

2.10 G. P. Irwin, . M. Krafft, P. C. Paris, and A. A. Wells, *Basic Aspects of Crack Growth and Fracture*, Naval Research Laboratory, Report 6598, Washington, D. C., November 1967.

2.11 American Institute of Steel Construction, *Specification for the Design, Fabrication, and Erection of Structural Steel for Buildings*, AISC, New York, February 1969.

2.12 T. R. Gurney, *Fatigue of Welded Structures*, Cambridge University Press, 1968.

2.13 J. R. Benjamin and C. A. Cornell, *Probability Statistics and Decision for Civil Engineers*, McGraw-Hill, New York, 1970.

2.14 American Society for Testing and Materials, *Fatigue Crack Propagation*, ASTM Special Technical Publication 415, Philadelphia, 1967.

2.15 P. C. Paris, "The Fracture Mechanics Approach to Fatigue," *Proceedings of the 10th Sagamore Conference*, Syracuse University Press, 1965, p. 107.

2.16 R. D. Stout, S. S. Tör and J. M. Ruzek, "The Effect of Fabrication Processes on Steels Used in Pressure Vessels," *Welding Journal*, Vol. 30, September 1951.

2.17 A. M. Freudenthal, "Safety, Reliability and Structural Design," *Transactions ASCE*, Vol. 127, 1962, Part II.

2.18 International Association for Bridge and Structural Engineering, *Symposium on Concept of Safety of Structures and Methods of Design*, Final Report, London, 1969.

2.19 N. C. Lind, *Deterministic Formats for the Probabilistic Design of Structures*, Department of Civil Engineering, Unversity of Waterloo, Ontario, Canada, November 1968.

2.20 J. M. Barsom and S. T. Rolfe, "Correlations between K_{IC} and Charpy V-Notch Test Results in the Transition-Temperature Range," *ASTM STP 466, Impact Testing of Metals*, 1970.

2.21 J. W. Fisher, K. H. Frank, M. A. Hirt, and B. M. McNamee, *Effect of Weldments on the Fatigue Strength of Steel Beams*, National Cooperative Highway Research Program Report 102, Washington, D.C., 1970.

2.22 J. Strating, "Loading Function," State of Art Report 3, Technical Committee 19, *Proceedings ASCE-IABSE International Conference on Tall Buildings*, 1973.

2.23 T. V. Galambos, "*Response Functions*," State of Art Report 4, Technical Committee 19, *Proceedings ASCE-IABSE International Conference on Tall Buildings*, 1973.

Chapter Three
Rivets

3.1 RIVET TYPES

Riveting is among the oldest methods of joining materials, dating back as far as the use of metals in construction practice.[1.8] Rivets were the most popular fasteners during the first half of this century, but their use has declined steadily since the introduction of the high-strength bolts. At the present time they are rarely used in field connections and face increasing competition from welding and high strength bolting in the shop.

Present specifications (1972) recognize two structural rivet steels namely ASTM A502 grade 1, low carbon rivet steel for general purposes, and ASTM A502 grade 2, a manganese rivet steel suitable for use with high-strength carbon and high-strength low alloy steel.[1.12]

The rivet heads are required to identify the grade and the manufacturer with appropriate marks. Markings can either be raised or depressed. For grade 1 the numeral 1 may be used at the manufacturer's option, but it is not required. The use of the numeral 2 to identify A502 grade 2 rivets is required.

Rivet steel strength is specified in terms of hardness requirements. The hardness requirements are applicable to the rivet bar stock of the full diameter as rolled. Figure 1.2 summarizes the hardness requirements for A502 rivet steels. There are no additional material requirements for strength or hardness in the driven condition.

3.2 INSTALLATION OF RIVETS

The riveting process consists of inserting the rivet in matching holes of the pieces to be joined and subsequently forming a head on the protruding end of the shank. The holes are generally $\frac{1}{16}$ in. greater than the nominal diameter of the undriven rivet. The head is formed by rapid forging with a pneumatic hammer or by continuous squeezing with a pressure riveter. The latter process is confined to use in shop practice, whereas pneumatic hammers are used in both shop and field riveting. In addition to forming the head, the diameter of the rivet is increased, resulting in a decreased hole clearance.

Most rivets are installed as hot rivets; that is, the rivet is heated to approximately 1800° before being installed. Some shop rivets are driven cold, a practice that is permissible if certain procedures are followed.

During the riveting process the enclosed plies are drawn together with installation bolts and by the rivet equipment. As the rivet cools, it shrinks and squeezes the connected plies together. A residual clamping force or internal tension results in the rivet. The magnitude of the residual clamping force depends on the joint stiffness and critical installation conditions such as driving and finishing temperature as well as the driving pressure. Measurements have shown that hot driven rivets can develop clamping forces that approach the yield load of a rivet. A considerable variation in clamping forces is generally observed.[3.3, 3.6, 3.7] Also, as the grip length is increased, the residual clamping force tends to increase.[3.7]

Residual clamping forces are also observed in cold driven rivets.[3.6] This results mainly from the elastic recovery of the gripped plies, after the riveter, which squeezed the plies together during the riveting process, is removed. Generally, the clamping force in cold formed rivets is small when compared to the clamping force in similar hot driven rivets.

The residual clamping force contributes to the slip resistance of the joint just as do high-strength bolts. However, the clamping force in the rivet is difficult to control, is not as great as developed by high strength bolts, and cannot be relied upon.

Upon cooling, rivets shrink lengthwise as well as diametrically. The amount of hole clearance that results also depends on how well the rivet filled the hole prior to shrinkage. Sawed sections of three hot formed, hand pneumatic driven rivets are shown in Fig. 3.1.[3.2] Studies have indicated that the holes are almost completely filled for relative short grip rivets. As the grip length is increased, clearances between rivet and plate material tend to increase. This tendency is due to the differences in working the material during driving.[3.2] Figure 3.1 shows some clearance for the longer grip rivets.

Installation of hot driven rivets involves many variables such as the initial or driving temperature, driving time, finishing temperature, and driving method. Over the years investigators have studied these factors, and where appropriate, these results are briefly discussed in the following section.

3.3 BEHAVIOR OF INDIVIDUAL FASTENERS

This section discusses briefly the behavior and strength of a single rivet subjected to either tension, shear, or combined tension and shear. Only typical test data are summarized in this chapter. No attempt was made to provide a comprehensive evaluation and statistical summary of the published test data.

3.3 Behavior of Individual Fasteners

3.3.1 Rivets Subjected to Tension

Typical stress strain curves for A502 grade 1 and grade 2 rivet steels are shown in Fig. 1.3. The tensile strength shown in Fig. 1.3 is about 60 ksi for grade 1 and 80 ksi for grade 2 rivets. These are typical of the values expected for undriven rivet materials.

The tensile strength of a driven rivet depends on the mechanical properties of the rivet material before driving and other factors related to the installation process. Studies have been made on the effect of driving temperature on the tensile strength. These tests indicated that varying the driving temperature between 1800 to 2300°F had little effect on the tensile strength.[3.1-3.3] It was also concluded on the basis of these test results that within practical limits, the soaking time, that is, the heating time of a rivet before driving, had a negligible effect on the ultimate strength.[3.2]

Driving generally increases the strength of rivets. For hot driven rivets it was observed that machine driving increased the rivet tensile strength by about 20%. The increase was about 10% for rivets driven by a pneumatic hammer. These same increases were observed when the tensile strength was determined from full size driven rivets and from specimens machined from

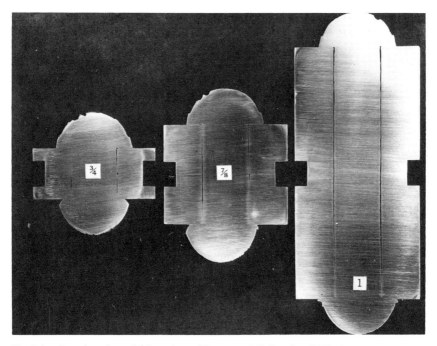

Fig. 3.1. Sawed sections of driven rivets. (Courtesy of University of Illinois.)

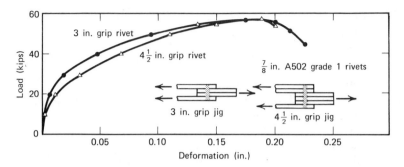

Fig. 3.2. Shear deformation curves for A502 grade 1 rivets.

driven rivets.[3.1] A considerable reduction in elongation was observed to accompany the increase in strength.

Tests also indicated that strain hardening of cold driven rivets resulted in an increase in strength.[3.1] Although only a few tests are available, they indicate that the increase in strength of cold driven rivets is at least equal to the increase in strength of similar hot driven rivets.[3.1, 3.4]

Most tension tests of driven rivets showed a tendency to decrease in strength as the grip length was increased. Two factors contribute to this observation. First, there is a greater "upsetting" effect, since the driving energy per unit volume for a short rivet is more favorable. Second, strength figures are based on the full hole area, which implies that the driven rivet completely fills the hole. As was noted in Fig. 3.1 this is not true for longer grip rivets as the gap increases with increasing grip length.[3.1, 3.2] For practical purposes the differences in strength of short and longer rivets is neglected.

It was reported in Ref. 3.2 that the residual clamping force in driven rivets has no influence on their strength. Yielding of the rivet minimizes the effect of the clamping force and does not affect the ultimate strength. A similar conclusion was reached for preloaded high strength bolts.[4.5-4.7]

3.3.2 Rivets Subjected to Shear

Many tests have been performed to evaluate the shear capacity of a rivet. It is common practice to express the shear strength of a rivet in terms of its tensile strength.[3.1, 3.2, 3.5] An average shear strength to tensile strength ratio of about 0.75 has been reported.[3.1, 3.2] The grade of the rivet material, as well as whether the test was performed on driven or undriven rivets, had little effect on this average value. Some of the data reported in Refs. 3.1

3.3 Behavior of Individual Fasteners

and 3.2 indicated that the shear to tensile strength ratio varied from 0.67 to 0.83. This wide variation is attributed to differences in testing methods, driving procedures, and differences in test specimens.

Figure 3.2 shows typical load-deformation curves for double shear tests on A502 grade 1 rivets.[3.8] Test results of two different grip lengths are shown. As expected, in the initial load stages, the longer rivet shows a larger deformation largely due to bending effects. The shear strength, however, was not affected.

Some data indicate a slight decrease in strength for rivets in single shear as compared to the double shear loading condition. This is caused by out of plane forces and secondary stresses on the rivet due to the inherent eccentricity of the applied load. In most single shear test joints, the rivet is not subjected to a pure shear load condition. When a single shear specimen is restrained so that no secondary stresses and out-of-plane deformations are introduced, the difference in the single and double shear strength is insignificant.[3.2]

Since driving a rivet increases its tensile strength, the shear strength is increased as well.[3.1, 3.2] If the average tensile strength of undriven A502 grade 1 and A502 grade 2 rivet materials is taken as 60 and 80 ksi, shear strengths between 45 and 60 ksi for grade 1 rivets and between 65 and 80 ksi for grade 2 rivets can be expected.

3.3.3 Rivets Subjected to Combined Tension and Shear

Tests have been performed to provide information regarding the strength and behavior of single rivets subjected to various combinations of tension and shear.[3.2] ASTM A141 rivets (comparable to A502 grade 1 rivets) were used for the study. The trends observed in this test series are believed to be applicable to other grades of rivets as well.

Among the test variables studied were variations in grip length, rivet diameter, driving procedure, and manufacturing process.[3.2] These variables did not have a significant influence on the results. Only the long grip rivets tended to show a decrease in strength. This was expected and was compatible with rivets subjected to shear alone.

As the loading condition changed from pure tension to pure shear, a significant decrease in deformation capacity was observed. This is illustrated in Fig. 3.3 where typical fractured rivets are shown for different shear to tension load ratios. Note that the character of the fracture and the deformation capacity changed substantially as the loading condition changed from shear to combined shear and tension and finally to tension.[3.2]

An elliptical interaction curve was fitted to the test results.[3.5] This

Fig. 3.3. Typical fractures at four shear tension ratios. (Courtesy of University of Illinois.)
Shear-tension ratio

1.0 : 0.0
1.0 : 0.577
0.577 : 1.0
0.0 : 1.0

defined the strength of rivets subjected to a combined tension and shear loading, as

$$\frac{x^2}{(0.75)^2} + y^2 = 1.0 \tag{3.1}$$

where x is the ratio of the shear stress on the shear plane to the tensile strength of the rivet (τ/σ_u) and y represents the ratio of the tensile stress to the tensile strength (σ/σ_u). The shear stress and tensile stress were determined on the basis of the applied loads. The tensile strength and shear strength were based on the rivet capacity when subjected to tension or shear only. The test results are compared with the elliptical interaction curve provided by Eq. 3.1 in Fig. 3.4 and show good agreement.

3.4 BASIS FOR DESIGN RECOMMENDATIONS

The behavior of individual rivets subjected to different types of loading conditions forms the basis for design recommendations. This section briefly summarizes rivet strength for the most significant loading conditions.

3.4 Basis for Design Recommendations

3.4.1 Rivets Subjected to Tension

The tensile capacity B_u of a rivet is equal to the product of the rivet cross-sectional area A_b and its tensile strength σ_u. The cross section is generally taken as the undriven cross section area of the rivet.[2.11] Hence

$$B_u = A_b \sigma_u \tag{3.2}$$

Depending on the type of rivet material, driving method, grip length, and such, σ_u may exceed the undriven rivet strength by 10 to 20%. A reasonable lower bound estimate of the rivet tensile capacity σ_u is 60 ksi for A502 grade 1 rivets and 80 ksi for A502 grade 2 rivets. Since ASTM specification do not specify tensile capacity these values can be used.

3.4.2 Rivets Subjected to Shear

The ratio of the shear strength τ_u to the tensile strength σ_u of a rivet was found to be independent of the rivet grade, installation procedure, diame-

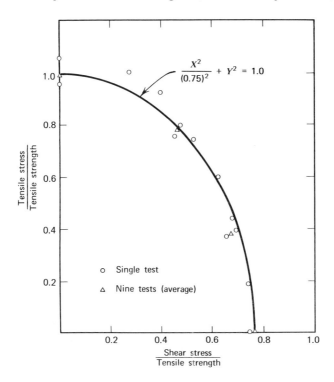

Fig. 3.4. Interaction curve for rivets under combined tension and shear.

ter, and grip length. Tests indicate the ratio to be about 0.75. Hence

$$\tau_u = 0.75\sigma_u \tag{3.3}$$

The shear resistance of a rivet is directly proportional to the available shear area and the number of critical shear planes. If a total of m critical shear planes pass through the rivet, the maximum shear resistance S_u of the rivet is equal to

$$S_u = 0.75mA_b\sigma_u \tag{3.4}$$

where A_b is the cross-section area of the undriven rivet.

3.4.3 Rivets Subjected to Combined Tension and Shear

The elliptical interaction curve given by Eq. 3.1 defines adequately the strength of rivets under combined tension and shear (see Fig. 3.4). Equation 3.1 relates the shear stress component to the critical tensile stress component. The product of ultimate stress and the undriven rivet area yields the critical shear and tensile load components for the rivet.

References

3.1 L. Schenker, C. G. Salmon, and B. G. Johnston, *Structural Steel Connections*, Department of Civil Engineering, University of Michigan, Ann Arbor, 1954.

3.2 W. H. Munse and H. C. Cox, *The Static Strength of Rivets Subjected to Combined Tension and Shear*, University of Illinois, Engineering Experiment Station Bulletin 437, Urbana, Illinois, 1956.

3.3 R. A. Hechtman, *A Study of the Effects of Heating and Driving Conditions on Hot-Driven Structural Steels Rivets*, Department of Civil Engineering, University of Illinois, Urbana, Illinois, 1948.

3.4 W. M. Wilson and W. A. Oliver, *Tension Tests of Rivets,* University of Illinois, Engineering Experiment Station Bulletin 210, Urbana, Illinois, 1930.

3.5 T. R. Higgins and W. H. Munse, "How Much Combined Stress Can a Rivet Take?," *Engineering News Record*, Vol. 23.

3.6 F. Baron and E. W. Larson, Jr., *The Effect of Grip on the Fatigue Strength of Riveted and Bolted Joints*, Research Report C110, The Technological Institute, Northwestern University, Evanston, Illinois, 1952.

3.7 F. Baron and E. W. Larson, Jr., *Comparative Behavior of Bolted and Riveted Joints*, Research Report C109, The Technological Institute, Northwestern University, Evanston, Illinois, 1952.

3.8 J. W. Fisher and N. Yoshida, "Large Bolted and Riveted Shingle Splices," *Journal of the Structural Division, ASCE*, Vol. 96, ST9, September 1969.

Chapter Four
Bolts

4.1 BOLT TYPES

The types of bolts used in connecting structural steel components in buildings and bridges can be categorized as follows (see Section 1.3):

1. Low carbon steel bolts and other fasteners ASTM A307, grade A.
2. High-strength medium carbon steel bolts ASTM A325, plain finish, weathering steel finish, or galvanized finish.
3. Alloy steel bolts, ASTM A490.
4. Special types of high strength bolts such as interference body bolts, swedge bolts, and other externally threaded fasteners or nuts with special locking devices, ASTM A449 and ASTM A354 grade BD.

ASTM A307 bolts require no head markings other than the manufacturers identification mark to appear on top of the head of this bolt.[1.10] These differ from high-strength A325 and alloy steel A490 bolts that require grade markings, as well as manufacturers identification. Also, A307 bolts are commonly made with both square and hex heads and are supplied with square and hex nuts respectively. Also, A307 nuts need not be identified. Figure 4.1 shows the two different types of A307 bolts and nuts.

In application, A307 bolts and nuts are tightened to some axial force to prevent movement of the connected members in the axial direction of the bolt. Proper tightening also prevents loosening of the nut. The actual force in the bolt is not closely controlled and may vary substantially from bolt to bolt. Because of small axial forces, little frictional resistance is developed and in most situations the bolt will slip into bearing. This results in shear stresses in the bolts and contact stresses at the points of bearing.

High-strength bolts are heat treated by quenching and tempering. Most widely used are A325 high-strength medium carbon steel bolts,[1.3] and A490 alloy steel bolts.[1.9] The A325 bolt is manufactured to one strength level in diameters from $\frac{1}{2}$ through 1 in. and to a slightly lower strength level over 1 to $1\frac{1}{2}$ in. On the other hand, A490 bolts are required to have a similar strength in all diameters from $\frac{1}{2}$ through $1\frac{1}{2}$ in.

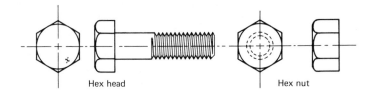
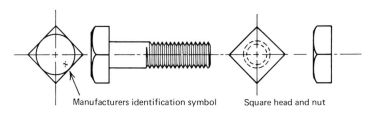

Fig. 4.1. A307 bolts.

The A325 and A490 bolts are heavy hex structural bolts used with plain hardened washers and heavy hex nuts as shown in Fig. 4.2.

The 1970a ASTM specification for A325 bolts provides for three types of high strength structural bolts.

Type 1. Bolts of medium carbon steel supplied in diameters from $\frac{1}{2}$ to $1\frac{1}{2}$ in. inclusive.

Type 2. Bolts of low carbon martensite steel supplied in diameters from $\frac{1}{2}$ to $1\frac{1}{2}$ in. inclusive.

Type 3. Bolts having atmospheric corrosion resistance and weathering characteristics comparable to that of A588 and A242 steels, supplied in diameters from $\frac{1}{2}$ to $1\frac{1}{2}$ in. inclusive.

The 1971 ASTM specification for A490 bolts changes the maximum diameter of structural bolts that may be manufactured under this specification from 4 to $1\frac{1}{2}$ in.

Heavy hex structural bolts manufactured to ASTM specification A325, type 1, 2, and 3 are identified on the top of the head by the legend A325, and the manufacturer's symbol, as is illustrated in Fig. 4.2. In addition type 1 bolts may be marked with three radial lines 120° apart, type 2 bolts must be marked with three radial lines 60° apart, and type 3 bolts must have the A325 underlined. On type 3 bolts, the manufacturer may add other distinguishing marks indicating that the bolt is of a weathering type.

A490 bolts are marked with the legend A490 and the manufacturer's symbol as illustrated in Fig. 4.2.

4.1 Bolt Types

Heavy hex A325 nuts and ASTM A563 grade C nuts are identified on at least one face by three equally spaced circumferential lines. Alternately, ASTM A194 grade 2 or grade 2H nuts and ASTM A563 grade D and grade DH nuts may be used. The A194 nuts are identified by the manufacturer's mark and the number 2 or 2H, respectively, and the A563 nuts are identified by the letters D or DH, respectively. Heavy hex nuts for A325 type 3 bolts are marked on at least one face with three equally spaced circumferential marks, and the number 3. In addition, the manufacturer may add other distinguishing marks indicating that the type 3 nut is of a weathering type.

Heavy hex nuts for use on A490 bolts may be ASTM A194 grade 2H or ASTM A563 grade DH and are identified by the legend 2H and by the manufacturer's mark and the legend DH, respectively.

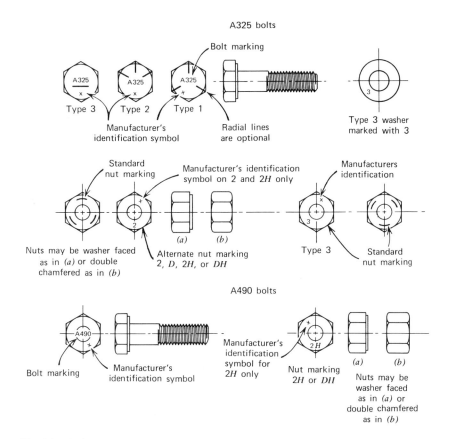

Fig. 4.2. Bolt markings for high-strength bolts.

Fig. 4.3. Interference body bolt. (Courtesy of Bethlehem Steel Corp.)

In addition to the standard A325 and A490 bolts $\frac{1}{2}$ through $1\frac{1}{2}$ in. diameter, short thread heavy head structural bolts above $1\frac{1}{2}$ in. diameter, and other types of fasteners and fastener components are available. These are covered by the general bolting specifications A449 and A354. Specification A449 covers externally threaded fastener products with similar mechanical properties to A325. The A354 grade BD covers externally threaded fastener parts that exhibit mechanical properties similar to A490.

Among the special types of fasteners or fastener components are the interference body bolts, swedge bolts, and nuts with locking devices. The interference body bolt (see Fig. 4.3) meets the strength requirements of the A325 bolt and has an axially ribbed shank that develops an interference fit in the hole and prevents excessive slip. A swedge bolt, shown in Fig. 4.4, consists of a fastener pin from medium carbon steel, and a locking collar of low carbon steel. The pin has a series of annular locking grooves, a breakneck groove, and pull grooves. The collar is cylindrical in shape and is

4.1 Bolt Types

swaged into the locking grooves in the tensioned pin by a hydraulically operated driving tool that engages the pull grooves on the pin and applies a tensile force to the fastener. After the collar is fully swaged into the locking grooves, the pin tail section breaks at the breakneck groove when its preload capacity is reached.

4.2 BEHAVIOR OF INDIVIDUAL FASTENERS

Connections are generally classified according to the manner of stressing the fastener (see Section 2.2), that is, tension, shear and combined tension, and shear. Typical examples of connections subjecting fasteners to shear are splices and gusset plates in trusses. Bolts in tension are common in hanger connections and in beam-to-column connections. Some beam-to-column connections may also subject the bolts to combined tension and shear. It is apparent that before a connection can be analyzed, the behavior of the component parts of the connection must be known. Therefore, the behavior of a single bolt subjected to the typical loading conditions of tension, shear and combined tension, and shear is discussed in this section.

4.2.1 Bolts Subjected to Tension

Since the behavior of a bolt subjected to an axial load is governed by the performance of its threaded part, load elongation characteristics of a bolt are more significant than the stress-strain relationship of the fastener metal itself.

In the 1970 ASTM specifications for high-strength bolts the minimum tensile strength and proof load are specified.[1.3, 1.9] The proof load is about

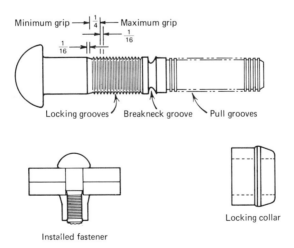

Fig. 4.4. High tensile swedge bolt.

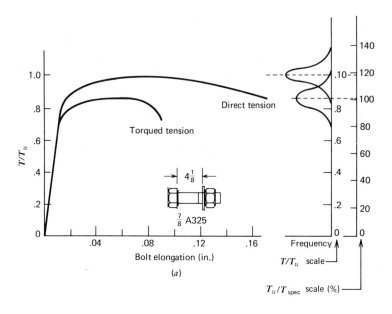

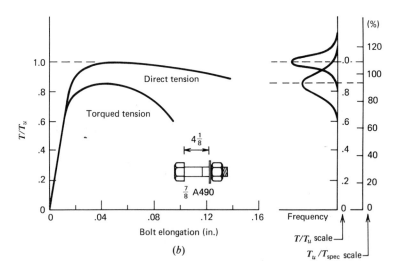

Fig. 4.5. (a) Load elongation relationship and frequency distributions of A325 bolts tested in torqued tension and direct tension; (b) A490 bolts.

4.2 Behavior of Individual Fasteners

equivalent to the yield strength of the bolt or the load causing 0.2% offset. To determine the actual mechanical properties of a bolt, ASTM requires a direct tension test of most sizes and lengths of full size bolts. In practice the bolt preload force is usually introduced by tightening the nut against the resistance of the connected material. As this torque is applied to the nut, the portion not resisted by friction between the nut and the gripped material is transmitted to the bolt and, due to friction between bolt and nut threading, induces torsional stresses into the shank. This tightening procedure results in a combined tension-torsional stress condition in the bolt. Therefore, the load-elongation relationship observed in a torqued tension test differs from the relationship obtained from a direct tension test. Torquing a bolt until failure results in a reduction in ultimate load as compared to the ultimate load determined from a direct tension test. Typical load elongation curves for direct tension as well as torqued tension tests

Fig. 4.6. Comparison of torqued tension and direct tension failures.

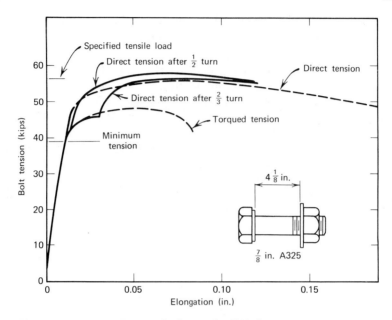

Fig. 4.7. Reserve tensile strength of torqued A325 bolts.

are shown in Fig. 4.5 for A325 bolts and A490 bolts, respectively. In torquing a bolt to failure a reduction in ultimate strength between 5 and 25% was experienced in tests on both A325 and A490 bolts.[4.1-4.3] The average reduction is equal to 15%. Frequency distributions of the ratio T/T_u for both A325 and A490 bolts are also shown in Fig. 4.5.

A bolt loaded to failure in direct tension has more deformation capacity than observed if the bolt is failed in torqued tension (see Fig. 4.6).[4.1-4.3] This is visible in the two specimens shown in Fig. 4.6. The differences in thread deformation and necking of the critical section in the threaded part of the bolts are apparent.

To determine whether minimum specified tensile requirements are met, specifications require direct tension tests on full size bolts, if the bolts are longer than 3 diameters, or if the nominal bolt diameter does not exceed $1\frac{1}{8}$ in. for A325 or $1\frac{1}{4}$ in. for A490 bolts. Bolts larger in diameter or shorter in length shall preferably be tested in full size; however, on long bolts tension tests on specimens machined from such bolts are allowed. Bolts shorter than three diameters need only meet minimum and maximum hardness requirements. Tests have illustrated that the actual tensile strength of production bolts exceed the minimum requirements considerably. An analysis of data obtained from tensile tests on bolts shows that A325 bolts in sizes $\frac{1}{2}$ through 1 in. exceed the minimum tensile strength required by 18%.

4.2 Behavior of Individual Fasteners

The standard deviation is equal to 4.5%. For larger diameter A325 bolts (1–1½ in.) the range of actual tensile strength exceeds the minimum by an even greater margin. A similar analysis of data obtained from tensile tests on A490 bolts shows an average actual strength 10% greater than the minimum prescribed. The standard variation is equal to 3.5%. Frequency distribution curves of the ratio T_u/T_spec are shown in Fig. 4.5a for A325 and in Fig. 4.5b for A490 bolts. Compared to the A325 the A490 bolts show a smaller margin beyond the specified tensile strength because specifications require the actual strength of A490 bolts within the range from 150 to 170 ksi whereas for A325 only a minimum strength is provided.

Loading a bolt in direct tension, after prestressing by tightening the nut, does not significantly decrease the ultimate tensile strength of the bolt, as illustrated in Figs. 4.7 and 4.8. The torsional stresses induced by torquing the bolt apparently have a negligible effect on the tensile strength of the bolt. This means that bolts installed by torquing can sustain direct tension loads without any apparent reduction in their ultimate tensile strength.[4.1, 4.2]

Mean load-elongation curves for 15 regular head ⅞-in. A325 bolts of various grips are plotted in Fig. 4.9.[4.2] The thickness of the gripped material varied from approximately 4¾ to 6¾ in. The length of thread under the nut varied from ¾ to 1 in. No systematic variation existed among the load-elongation relationships for the different grip conditions. Since the length of thread under the nut is relatively constant, grip has no appreciable effect on these relationships. The behavior shown in Fig. 4.9 for the direct tension

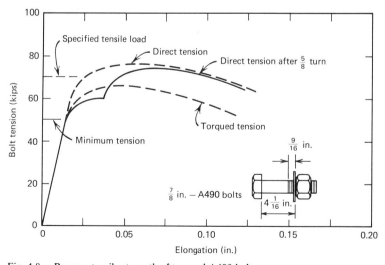

Fig. 4.8. Reserve tensile strength of torqued A490 bolts.

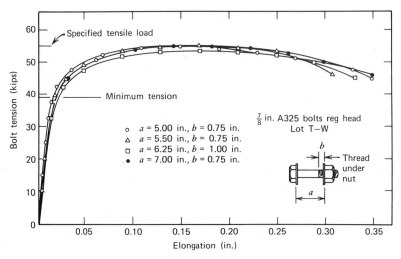

Fig. 4.9. Effect of grip length, direct tension.

test was also observed during torqued tension tests. With shorter grip lengths the effect of bolt length is more pronounced.

Figure 4.9 shows also, that within the elastic range, the elongation increases slightly with an increase in grip. As the load is increased beyond the elastic limit, the threaded part, which is approximately of uniform length, behaves plastically while the shank remains essentially elastic. Hence when there is a specific amount of thread under the nut, grip length has little effect on the load elongation relationship beyond the proportional limit. For short bolts nearly all deformation occurs in the threaded length and causes a decrease in rotational capacity.

Heavy head bolts demonstrated similar behavior for grips ranging from 4 to 8 in. and with thread lengths under the nut from $\frac{1}{8}$ to $\frac{3}{4}$ in. Similar observations have also been made on tests on A490 bolts.[4.1, 4.3]

Since most of the elongation occurs in the threads, the length of thread between the thread run-out and the face of the nut will affect the load elongation relationship. The heavy head bolt has a short thread length whereas the regular head bolt has the normal ASA thread length specified by ANSI standards. As a result, for a given thickness of gripped material the heavy head bolt shows a decrease in deformation capacity, as illustrated in Fig. 4.10.[4.2]

4.2.2 Bolts Subjected to Shear

Shear-deformation relationships have been obtained by subjecting fasteners to shear induced by plates either in tension or compression. Typical results

4.2 Behavior of Individual Fasteners

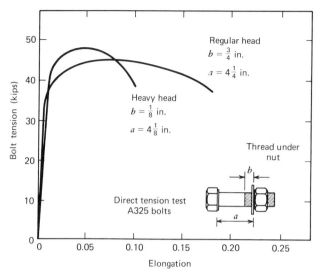

Fig. 4.10. Comparison of regular and heavy head A325 bolts.

of shear tests on A325 and A490 bolts are shown in Fig. 4.11. As expected, an increase in tensile strength increases the shear strength. A slight decrease in deformation capacity is evident as the strength of the bolt increases.[4.4]

The shear strength is influenced by the type of test. The fastener can be subjected to shear by plates in tension or compression as illustrated in Fig. 4.12. The influence of the type of test on the bolt shear and deformation

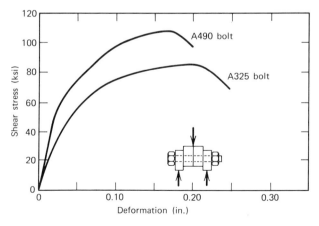

Fig. 4.11. Typical shear deformation curves for A325 and A490 bolts.

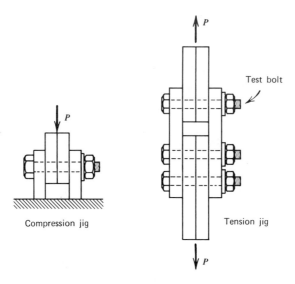

Fig. 4.12. Schematic testing jigs for single bolts.

capacity is illustrated in Fig. 4.13 where typical shear stress-deformation curves are compared for bolts from the same lot that were tested in both tension and compression jigs.[4.4] Test results show that the shear strength of bolts tested in A440 steel tension jigs is 6 to 13% lower than bolts tested in A440 steel compression jigs.[4.4] The same trend was observed in constructional alloy steel jigs where the reduction in shear strength of similar bolts varied from 8 to 13%. The average shear strengths for A325 and A490* bolts tested in tension jigs were 80.1 and 101.1 ksi, respectively. This corresponds to about 62% of the bolts' tensile strengths. The same bolt grades tested in compression jigs yielded shear strengths of 86.5 and 113.7 ksi, respectively (68% of the bolt tensile strength).[4.4]

The lower shear strength of a bolt observed in a tension type shear test as compared to a compression type test (see Fig. 4.13) is the result of lap plate prying action, a phenomenon that tends to bend the lap plates of the tension jig outward.[4.4, 4.25] Because of the uneven bearing deformations of the test bolt, the resisting force does not act at the centerline of the lap plate, which produces a moment that tends to bend the lap plate away from the main plate. This moment causes tensile forces in the bolt.

Catenary action, resulting from bending in bolts, may also contribute to the increase in bolt tension near ultimate load. However, it is believed that

---

* Actually, A354 grade BD bolts were used instead of A490 bolts because of their similarity in mechanical properties. At the time of the tests the A490 bolt was under development.

4.2 Behavior of Individual Fasteners

this effect is small in comparison to the tension induced by lap plate prying action.[4.25] In any case, the catenary action is present in both the tension and compression jigs.

The tension jig is recommended as the better testing device to be used so as to obtain a lower bound shear strength. Bolts in tension splices are subjected to shear in a similar manner. Also, the tension-jig shear test yields the most consistent test results.

An examination of available test data indicates that the ratio of the shear strength to the tensile strengh is independent of the bolt grade as illustrated in Fig. 4.14. The shear strength is plotted versus the tensile strength for various lots of A325 and A490 bolts. The average shear strength is approximately 62% of the tensile strength.

The variance of the ratio of the shear strength to tensile strength, as obtained from single bolt tension shear jigs, is shown in Fig. 4.15. A frequency curve of the ratio of shear strength to tensile strength was developed from test data acquired at the University of Illinois and Lehigh University. The average value is equal to 0.62 with a standard deviation of .033.

Tests on bolted joints indicated that the initial clamping force had no significant effect on the ultimate shear strength.[4.5-4.7] To illustrate this a number of tests were performed on A325 and A490 bolts torqued to var-

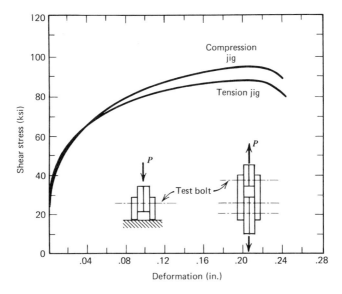

Fig. 4.13. Typical shear deformation curves A354 BC bolts tested in A440 steel tension and compression jigs.

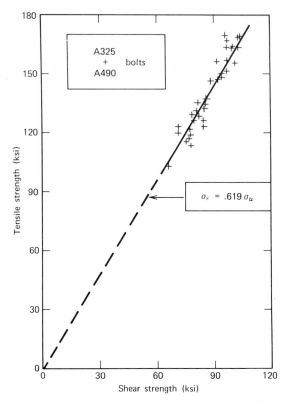

Fig. 4.14. Shear strength plotted versus tensile strength. *Note.* Each point represents the average values of a specific bolt lot. The shear strength is computed on the relevant area, depending on the location of the shear plane.

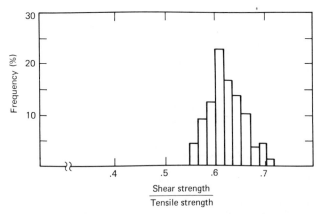

Fig. 4.15. Frequency distribution of ratio of shear strength to tensile strength for A325 and A490 bolts. Number of tests, 142; average value, 0.625; standard deviation, 0.033.

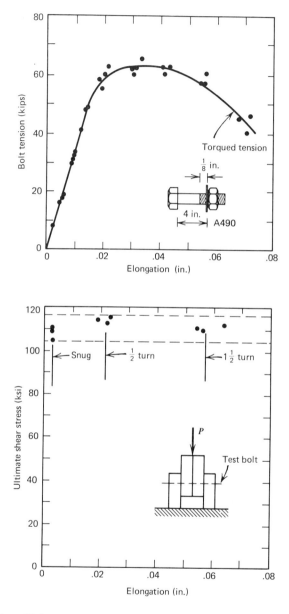

Fig. 4.16. Effect of bolt preload on shear strength of A490 bolts.

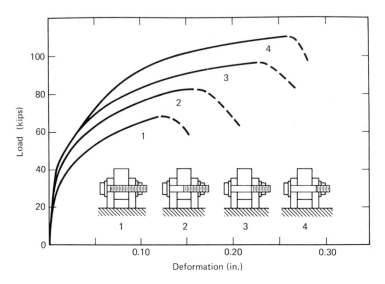

Fig. 4.17. Shear-deformation curves for different failure planes.

ious degrees of tightness and then tested to failure in double shear. The results of tests with A490 bolts are shown in Fig. 4.16.[4.4] The lower portion shows the relationship between bolt shear strength and the initial bolt elongation after installation. The bolt preload was determined from measured elongations and the torqued tension relationship given in the upper portion of Fig. 4.16. The figure confirms that no significant variation of shear strength occurred when the initial bolt preload was varied.

A number of factors are responsible for this fact. Measurements of the internal tension in bolts installed in joints have indicated that at the ultimate load level there is little clamping force left in the bolt.[4.6, 4.7, 4.25] Depending on the plastic deformations in the plates, prying action may induce axial bolt tension. In most practical situations, however, the tensile stress induced by prying action will be considerably below the yield stress of the bolt; therefore, it has only a minor influence. Studies of bolts under combined tension and shear have shown that tensile stresses equal to 20 to 30% of the tensile strength do not significantly affect the shear strength of the bolt.[4.8]

Furthermore, it has to be noted that the critical shear plane is often through the bolt shank. Because of the increased area, the tensile stress in the shank is significantly lower than the tensile stress in the threaded portion of the bolt; consequently, its influence on the shear strength is greatly reduced.

The shear resistance of high-strength bolts is directly proportional to the

4.2 Behavior of Individual Fasteners

available shear area. The available shear area in the threaded part of a bolt is equal to the root area and is less than the area of the bolt shank. For most commonly used bolts the root area is about 75% of the nominal area. The influence of the shear plane location on the load deformation characteristics of A325 and A490 bolts is reported in Ref. 4.4. Figure 4.17 shows the influence of the shear plane location on the load displacement behavior of A325 bolts. When both shear planes passed through the bolt shanks, the shear load and deformation capacity were maximized. When both shear planes passed through the threaded portion, the lowest shear load and deformation capacity were obtained. All available tests indicate that the shear resistance of both A325 and A490 bolts is governed by the available shear area. The shear strength was unaffected by the shear plane location.

4.2.3 Bolts Subjected to Combined Tension and Shear

To provide information regarding the strength and behavior characterisitcs of single high-strength bolts subjected to various combinations of tension and shear, tests were performed at the University of Illinois.[4.8] Two types of high-strength bolts, A325 and A354 grade BD, were used in the investigation. Since the mechanical properties of A354 grade BD and A490 bolts are nearly identical, the data are directly applicable to A490 as well as to A354 BD bolts.

Certain other factors that might influence the performance of high-strength bolts under combined loadings of tension and shear were also examined in the test program. These included (a) bolt grip length, (b) bolt diameter, (c) type of bolt, and (d) type of material gripped by the bolt. In addition, the influence of the location of the shear planes was examined.

The Illinois tests indicated that an increase in bolt grip tends to increase the ultimate load of a bolt subjected to combined tension and shear. This increase in resistance is mainly caused by the greater bending that can develop in a long bolt as compared to a short grip bolt. At high loads the short grip bolt presented a circular shear area, whereas the long grip bolt, because of bending, presented an elliptical cross-section with a larger shear area.

It was concluded, however, that neither the test block material nor the bolt diameter had a significant effect on the ultimate load capacity of the bolt.

Figure 4.18 summarizes test results of bolts subjected to combined tension and shear.[4.8] The tensile strength (in kilopounds per square inch) was used to nondimensionalize the shear and tensile stresses due to the shear and tensile components of the load. The tensile stress was computed on the basis of the stress area, whereas the shear stress is dependent on the location of the shear plane. An elliptical interaction curve was used to represent

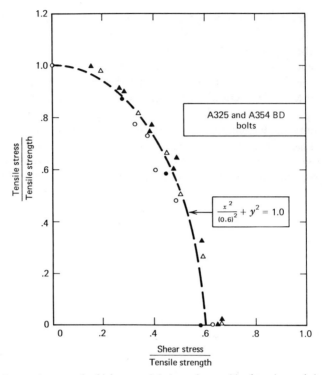

Fig. 4.18. Interaction curve for high-strength bolts under combined tension and shear.

	A325	A354BD
Threads in shear plane	△	○
Shank in shear plane	▲	●

the behavior of high-strength bolts under combined tension and shear,

$$\frac{x^2}{(0.62)^2} + y^2 = 1.0 \tag{4.1}$$

where x is the ratio of the shear stress on the shear plane to the tensile strength and y represents the ratio of the tensile stress to the tensile strength (both computed on the stress area). Figure 4.18 also indicates that neither the bolt grade nor the location of the shear plane influence the ultimate x/y ratio. This is compatible with the behavior of bolts in pure shear.

4.3 INSTALLATION OF HIGH-STRENGTH BOLTS

To provide the desired level of preload, high-strength structural bolts must be tightened so that the resulting bolt tension is at least 70% of the mini-

4.3 Installation of High-Strength Bolts

mum required tensile strength. The required minimum bolt tension is given in the RCRBSJ specifications, as listed in Table 4.1.

When the high-strength bolt was first introduced, installation was primarily dependent on torque control. Approximate torque values were suggested for use to obtain the minimum specified bolt tension. Tests performed by Maney,[4.12] and later by Pauw and Howard,[4.13] showed the great variability of the torque-tension relationship. Bolts from the same lot yielded extreme values of bolt tension $\pm 30\%$ from the mean tension desired. The average variation was in general $\pm 10\%$. This variance was mainly caused by the variability of the thread conditions, surface conditions under the nut, lubrication, and other factors that cause energy dissipation without inducing tension in the bolt. Experience in field use of high-strength bolts confirmed the erratic nature of the torque-tension relationship.

Current specifications permit high-strength bolts to be tightened by using calibrated wrenches, by the turn-of-nut method, or by use of direct tension indicators.[1.4] The last two procedures depend on strain or displacement control as contrasted to the torque control of the calibrated wrench method.

In the calibrated wrench method the wrench is calibrated or adjusted to "stall" when the desired tension is reached. In practice, several bolts of the

Table 4.1. Fastener Tension

Bolt Size (in.)	Minimum Fastener Tension[a] in Thousands of Pounds (kips)	
	A325 Bolts	A490 Bolts
$\frac{1}{2}$	12	15
$\frac{5}{8}$	19	24
$\frac{3}{4}$	28	35
$\frac{7}{8}$	39	49
1	51	64
$1\frac{1}{8}$	56	80
$1\frac{1}{4}$	71	102
$1\frac{3}{8}$	85	121
$1\frac{1}{2}$	103	148
Over $1\frac{1}{2}$		$0.7 \times$ T.S.

[a] Equal to 70% of specified minimum tensile strengths of bolts, rounded off to the nearest kip.

lot to be installed are tightened in a calibrating device that directly reads the tension in the bolt. The wrench is adjusted to stall at bolt tensions which are a minimum of 5% greater than the required preload. To minimize the variation in friction between the underside of the turned surface and the gripped material, hardened washers must be placed under the element turned in tightening. A minimum of three bolts of each diameter must be tightened at least once each working day in a calibrating device capable of indicating actual bolt tensions. This check should also be performed each time significant changes are made in the equipment or when a significant difference is noted in the surface conditions of the bolts, nuts, or washers.

To overcome the variability of torque control, efforts were made to develop a more reliable tightening procedure. The American Association of Railroads (AAR), faced with the problem of tightening bolts in remote areas without power tools, conducted a large number of tests to determine if the turn-of-nut could be used as a means of controlling bolt tension.[4.14, 4.15] These tests led to the conclusion that one turn from a finger tight position produced the desired bolt tension. In 1955 the RCRBSJ adopted one turn of the nut from hand tight position as an alternative method of installation.

Experience with the one full turn method indicated that it was impractical to use finger or hand tightness as a reliable point for starting the one turn. Because of out-of-flatness, thread imperfections, and dirt accumulation, it was difficult and time consuming to determine the hand tight position.

Bethlehem Steel Corporation developed a modified "turn-of-nut" method, using the AAR studies and additional tests of their own.[4.16, 4.17] This method called for running the nut up to a snug position using an impact wrench rather than the finger tight condition. From the snug position the nut was given an additional $\frac{1}{2}$ or $\frac{3}{4}$ turn, depending on the length of the bolt. The snug condition was defined as the point at which the wrench started to impact. This occurred when the turning of the nut was resisted by friction between the face of the nut and the surface of the steel. Snug-tightening the bolts induces small clamping forces in the bolts. In general at the snug-tight condition, the bolt clamping forces vary considerably within the elastic range. This is illustrated in Fig. 4.19 where the range of bolt clamping forces and bolt elongations at the snug tight condition is shown for $\frac{7}{8}$ in. A325 bolts installed in an A440 steel test joint. The average clamping force at the snug tight condition was equal to about 26 kip. The bolts in this test joint were snug tightened by means of an impact wrench. This modified turn-of-nut method was eventually incorporated into the 1960 council specifications.

4.3 Installation of High-Strength Bolts

For bolts equal or greater than ¾-in. diameter, snug position provided by an impact wrench is approximately equal to the tightness attained by the full effort of a man using an ordinary spud wrench. For bolts smaller than ¾ in. in diameter, a man's full effort may deform the bolt into the inelastic range. This is not desirable, because it results in a severe reduction in nut-rotation capacity. Therefore, small diameter bolts (½ and ⅝ in.) must be snug tightened using different criteria. For tightening such bolts a wrench with a smaller length than usual could be used or a small torque wrench can be used to provide a suitable snug-tight condition.

Controlling tension by the turn-of-nut method is primarily a strain control, and the effectiveness of the method depends on the starting point and

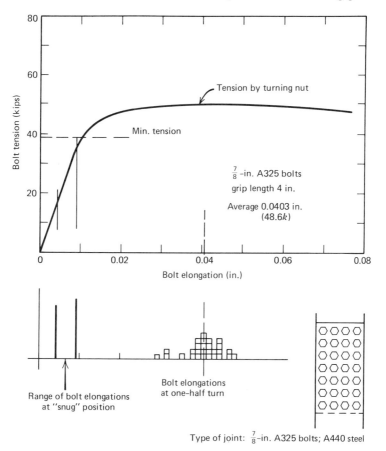

Fig. 4.19. Bolt elongation "snug" and after additional one-half turn of nut. Type of joint: ⅞-in. A325 bolts; A440 steel.

Table 4.2. Nut Rotation from Snug-Tight Condition[a]

Disposition of Outer Faces of Bolted Parts		
Both faces normal to bolt axis, or one face normal to axis and other face sloped not more than 1:20 (bevel washer not used)		Both faces sloped not more than 1:20 from normal to bolt axis (bevel washers not used)
Bolt length[b] not exceeding 8 diameters or 8 in.	Bolt length[b] exceeding 8 diameters or 8 in.	For all length of bolts
$\frac{1}{2}$ turn	$\frac{2}{3}$ turn	$\frac{3}{4}$ turn

[a] Nut rotation is rotation relative to bolt regardless of the element (nut or bolt) being turned. Tolerance on rotation: 30° over or under. For coarse thread heavy hex structural bolts of all sizes and length and heavy hex semi-finished nuts.

[b] Bolt length is measured from underside of head to extreme end of point.

accuracy of the rotational measurements. If these factors are carefully controlled, desired tensions can be obtained with accuracy in both the elastic and inelastic ranges. In the inelastic region, the load elongation curve for a bolt is relatively flat so that variations in snug result in only minor tension variations. The latter has been illustrated by studies on test joints. A typical example is shown in Fig. 4.19. Extensive research in this area has indicated that the half turn of nut was adequate for all lengths of A325 bolts.[4.2, 4.5-4.7, 4.9] Based on this experience the 1962 edition of the council specifications required only one-half turn regardless the bolt length.

In 1964 the RCRBSJ incorporated the A490 bolt into its specification. In making the council specification applicable to both A325 and A490 bolts the turn-of-nut method was modified again. Tests of A490 bolts had indicated that when the grip length was increased to about eight times the bolt diameter, a somewhat greater nut rotation was needed to reach the required minimum bolt tension. Therefore, additional rotation for all bolts is required (see Table 4.2). Although the additional rotation was not needed for A325 bolts, the two-thirds turn provision has been applied to the A325 bolts as well, in the interest of uniformity in field practice.

Calibration tests of A325 bolts with grips more than four diameters or 4 in. showed that the one-half turn of the nut rotation produced consistent bolt tensions in the inelastic range.[4.2] These tests also showed a sufficient margin of safety against fracture by excessive nut rotation. Bolts with grips

4.3 Installation of High-Strength Bolts

more than 4 in. or four diameters and short thread length under the nut can be given one-half turn of the nut and have sufficient deformation capacity to sustain two additional half turns before failure. Bolts with long thread lengths in the grip can sustain three to five additional half turns, as illustrated in Fig. 4.20. Similar tests on A490 bolts are compared with the results of A325 bolts in Fig. 4.21. A325 and A490 bolts gave substantially the same load-nut rotation relationships up to the elastic limit.[4.1, 4.3, 4.9] At one-half turn from the snug position the A490 bolts provided approximately 20% greater load than A325 bolts, because of the increased strength of the A490 bolt. However, the higher strength of the A490 bolts results in a small decrease in nut rotation capacity as compared to the A325 bolt. Therefore, the rotational factor of safety against twisting off is similar to the A325 bolt.

Studies on short bolts (less than four diameters or 4 in., whichever is greater) have shown that their rotational factor of safety against twisting off is less than two. Care should be exercised in their installation so that they are not "overtightened." One-third turn of the nut provides a bolt preload above the minimum tension.

The latest edition of the council specification permits high-strength bolts to be tightened by use of a direct tension indicator, provided it can be demonstrated that both have been tightened in accordance with Table 4.1.[1.4] Several systems which fall into this category have been developed

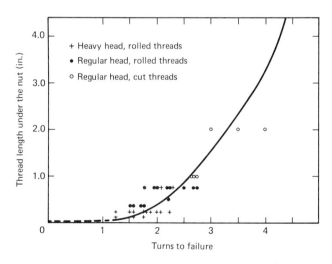

Fig. 4.20. Effect of thread length on rotation capacity of A325 bolts.

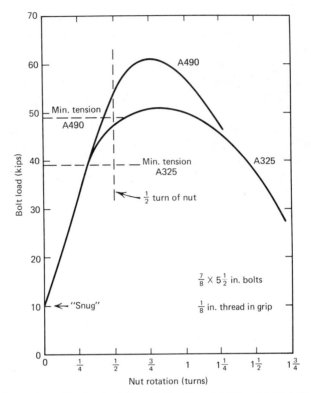

Fig. 4.21. Comparison of load-nut rotation relationships of A490 and A325 bolts.

and are commercially available. The swedge bolt[4.29] and load indicating washers[4.28] fall into this category.

Specifications require that the slope of surfaces of bolted parts in contact with the bolt head or nut shall not exceed 1:20 with respect to a plane normal to the bolt axis. Research carried out at the University of Illinois determined the influence of beveled surfaces (1:20 slope) when bevel washers were omitted.[4.9] A325 bolts are ductile enough to deform to this slope. Greater slopes are undesirable as they affect both strength and ductility.

From these tests it was concluded that the inclusion of bolted connections with a 1:20 slope in the grip and without beveled washers requires additional nut rotation to ensure that tightening will achieve the required minimum tension.[4.9] No additional rotation is necessary for one beveled surface (although the resulting bolt force may be near the minimum tension); however, with beveled surfaces under both the head and the nut an additional one-fourth turn should be used.

4.4 RELAXATION

Due to the high stress level in the threaded part of an installed bolt, some relaxation will occur that could affect the bolt performance. To evaluate the influence of this relaxation, studies were performed on assemblies of A325 and A354 grade BD bolts in A7 steel.[4.9] The bolts were tightened by turning the nut against the gripped material. The bolt tension versus time was registered throughout the study.

From these tests it was evident that immediately upon completion of the torquing there was a 2 to 11% drop in load. The average loss was 5% of the maximum registered bolt tension. This drop in bolt tension is believed to result from the elastic recovery which takes place when the wrench is removed. Creep and yielding in the bolt due to the high stress level at the root of the threads might result in a minor relaxation as well.

The grip length as well as the number of plies are believed among the factors that influence the amount of bolt relaxation. Although no experimental data are available, it seems reasonable to expect an increase in bolt force relaxation as the grip length is decreased. Similarly, increasing the number of plies for a constant grip length might lead to an increase in bolt relaxation. The loss in bolt preload can be large for very short grip bolts (i.e., $\frac{1}{2}$ to 1-in. grips).

Relaxation tests on A325 and A354 BD bolts showed an additional 4% loss in bolt tension after 21 days as compared with the bolt tension measured 1 min after torquing.[4.9] Ninety percent of this loss occurred during the first day. During the remaining 20 days the rate of change in bolt load decreased in an exponential manner.

Relaxation studies on assemblies with high-strength bolts were performed in Japan and showed similar results.[4.10] By extrapolating the test data it was concluded that the relaxation after 100,000 hr (11.4 years) could be estimated at about 6% of the bolt load immediately after tightening.

The relaxation characteristics of assemblies of galvanized plates and bolts were found to be about twice as great as plain bolts and connected material.[4.19] The amount of relaxation appeared to be related to the thickness of the galvanized coating. It was concluded that the increased bolt relaxation occurred because of the creep or flow of the zinc coating under sustained high clamping pressures. Like plain ungalvanized bolts, the galvanized bolts experienced most of the creep and relaxation immediately upon completion of the tightening process.

Based on tests performed at Lehigh University it was concluded that within certain limits oversize or slotted holes do not significantly affect the losses in bolt tension with time, following installation.[4.26] The loss in ten-

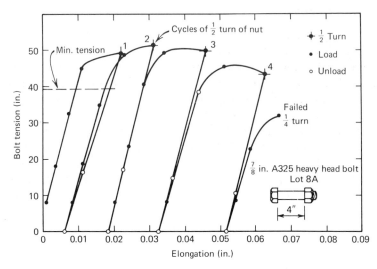

Fig. 4.22. Repeated installation of A325 bolts.

sion was about 8% of the initial preload. A more detailed discussion on this is given in Chapter 9.

4.5 REUSE OF HIGH-STRENGTH BOLTS

Since the turn-of-nut method often induced a bolt tension that exceeds the elastic limit of the threaded portion, repeated tightening of high-strength bolts may be undesirable. Tests were performed to examine the behavior of high-strength bolts under alternately torquing one-half turn, loosening, and retorquing.[4.1, 4.2] The record of one such test on a A325 bolt is summarized in Fig. 4.22. It is apparent that the cumulative plastic deformations caused a decrease in the A325 bolt deformation capacity after each succeeding one-half turn. However, A325 bolts can be reused once or twice, providing that proper control on the number of reuses can be established.

Whereas the as-received black A325 bolts generally do have adequate nut rotation capacity to allow for a limited reuse, reuse of coated A325 bolts is not recommended. Tests have indicated that the nut rotation capacity of a bolt is generally reduced by providing a coating (see Section 4.6).[4.19, 4.27] Therefore, unless experimental data indicate otherwise, reuse of coated A325 bolts should not be permitted.

Figure 4.23 shows typical results of one lot of A490 bolts repeatedly installed with threads as received. Note that the minimum required tension only was achieved during the first and second cycle. Subsequent cycles showed a sharp decrease in induced bolt tension. Test results have indicated

4.5 Reuse of High-Strength Bolts

that bolts from the same lot when waxed had considerably improved characteristics.[4.1] However, whether the threads were waxed or as-received, a marked increase in installation time was noted for successive cycles. The behavior of A490 bolts under repeated torquing seems to be more critical than A325 bolts. Therefore, reuse of A490 bolts is not recommended.

4.6 GALVANIZED BOLTS AND NUTS

At the present time a wide range of structures are being treated with a protective surface coating to prevent corrosion and reduce maintenance costs. Galvanizing is a widely used procedure and provides an excellent corrosion resistant protection.

The behavior of galvanized bolts may differ from the behavior of normal, uncoated high-strength bolts.[4.18, 4.19] This difference in behavior is primarily caused by the zinc layer on the bolt threads, galling in the threads, and seizing when the bolt is tightened. Occasionally this makes it difficult to reach the desired bolt tension without experiencing a premature torsional failure of the bolt.

The zinc coating on the surface of a bolt does not affect the bolt static strength properties. Calibration studies showed that the tensile strength, as determined from a direct tension test, as well as the shear strength of the bolt were not affected by the galvanizing process.[4.18, 4.19] However, if bolt tension is induced by turning the nut against the gripped material, unlubricated galvanized bolts experienced a greater reduction in the maximum bolt tension as compared with torqued ungalvanized bolts or properly lubricated galvanized bolts. This reduction was up to 25% more than plain black bolts, depending on the thread conditions and thickness of the zinc layer.

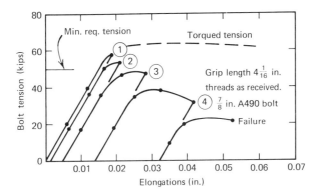

Fig. 4.23. Repeated installation of A490 bolts.

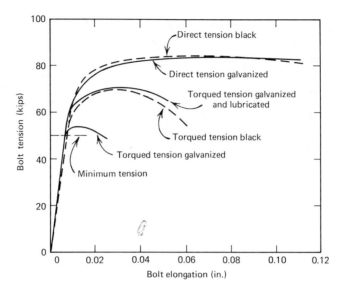

Fig. 4.24. Comparison load elongation relationships between 1 inch black and galvanized A325 bolts.

Besides this reduction in torqued tension strength, the added frictional resistance on the threads of the galvanized bolts caused a considerable decrease in ductility, as illustrated in Fig. 4.24. This effect of high frictional resistance can be reduced substantially by employing lubricants on the threads of galvanized bolts. Tests indicated no appreciable difference in the torqued tensile strength of plain bolts as-received and galvanized bolts lubricated with either beeswax, cetyl alcohol, or commercial wax.[4.27, 4.11] Some reduction in ductility of the galvanized bolts was observed. Calibration tests performed on galvanized A490 bolts showed results similar to the results of A325 bolts.[4.18]

A high tendency for stripping-type failures was observed in torqued tension tests of galvanized high-strength bolts.[4.19] This can be attributed to several factors. As the bolt is torqued, the threaded section within the grip necks down and the nut spreads. This along with the overlapping of the nut that is necessary for galvanizing may cause an excessive disengagement of some of the threads in the nut and increase the chance for stripping failures. To reduce the possibility of an undesirable stripping failure, harder nuts should be used for galvanized bolts (nuts of quality DH or 2H).

Although galvanizing does provide an excellent protection against corrosion of the bolt, it may increase its susceptibility to stress corrosion and

4.6 Galvanized Bolts and Nuts

hydrogen stress cracking. This applies especially to galvanized A490 bolts. Therefore, it was concluded that galvanized A490 bolts should not be used in structures.[4.23, 4.24]

4.7 USE OF WASHERS

Originally the high-strength structural bolt assembly included a bolt with a nut and two hardened washers. The washers were thought necessary to serve the following purposes:

1. To protect the outer surface of the connected material from damage or galling as the bolt or nut was torqued or turned.
2. To assist in maintaining a high clamping force in the bolt assembly.
3. To provide surfaces of consistent hardness so that the variation in the torque-tension relationship could be minimized.

When the turn-of-nut method for tightening high-strength bolts was adopted, a procedure was introduced which provided a means of obtaining the required bolt tension without reliance upon torque-tension control. Hence it was desirable to determine whether hardened washers were needed in the bolt assembly. Tests showed that a hardened washer was not needed to prevent minor bolt relaxation resulting from the high stress concentration under the bolt head or nut of A325 bolts.[4.9] It was also concluded that any galling that may take place when nuts for A325 bolts are tightened directly against the connected parts is not detrimental to the static or fatigue strength of the joint.

As a result of these findings, the council specifications in general do not require the use of washers when A325 bolts are installed by the turn-of-nut method. Where bolts are tightened by the calibrated wrench method (i.e., torque control), a washer is required under the turned element (the nut or bolt head). Washers are also required when A490 bolts connect A36 steel parts, to reduce galling and brinelling of these parts. In high-strength steel they are only required to prevent galling of the turned element.

When bolts pass through a beam or channel flange that has a sloping interface, a bevel washer is often used to compensate for the lack of parallelism. Specifications require the use of beveled washers when an outer face has a slope greater than 1:20. A325 bolts are ductile enough to deform to this slope.[4.9] Greater slopes are undesirable as they affect both strength and ductility.

With a 1:20 slope in the gripped material, bolts require additional nut rotation to ensure that tightening will achieve the minimum tension.

4.8 CORROSION AND EMBRITTLEMENT

Under certain conditions, corrosive environments may be detrimental to the serviceability of coated high-strength bolts subjected to sustained stresses. Hydrogen stress cracking as well as stress corrosion may cause delayed, "brittle" fractures of high-strength bolts. Although both processes have been studied extensively, no completely acceptable mechanism for explaining either phenomenon has been developed.

In many respects the two fracture mechanisms have a number of similarities. Both may cause delayed, brittle-type fractures of bolts. However, there appear to be significant differences. For example, stress corrosion at least in part involves electrochemical dissolution of metal along active sites under the influence of tensile stress. Hydrogen stress cracking occurs as the result of a combination of hydrogen in the metal lattice and tensile stress. The hydrogen produces a hard martensite structure that is susceptible to cracking. Atomic hydrogen absorption by the steel is necessary for this type of failure to occur. Since corrosion frequently is accompanied by the liberation of atomic hydrogen, hydrogen-stress cracking may occur in corrosive environments. However, in many situations a combination of both fracture patterns develops.

Laboratory tests have shown that both phenomena influence the life of high-strength bolts.[4.22-4.24] The behavior of A325 as well as A490 bolts under different environmental conditions was studied. From these test results, it became apparent that the higher the strength of the steel, the more sensitive the material becomes to both stress corrosion and hydrogen stress cracking. The study indicated a high susceptibility of galvanized A490 bolts to hydrogen stress cracking. It was concluded that this was caused by a break in the zinc film which promoted the entry of atomic hydrogen into the metal. If there were no breaks in the coating, failures were not likely to occur. The study also indicated the desirability of limiting the hardness of A490 bolts. Several uncoated bolts were observed to fail when high hardness and strength were present. Because of this observation, the maximum tensile strength was decreased by ASTM.

On the basis of these tests it was concluded that properly processed black and galvanized A325 bolts, heat treated within presently specified hardness limits, will behave satisfactorily with regard to hydrogen stress and stress corrosion cracking in most corrosive environments.[4.23] Particular attention should be given to the preparation of the bolts for galvanizing. Improper pickling procedures could induce hydrogen embrittlement. It was further concluded that galvanized A490 bolts should not be used in structures. The tests did indicate that black A490 bolts can be used without problems from "brittle" failures in most environments.

4.9 EFFECT OF NUT STRENGTH

The behavior of bolt assemblies may vary when tightened to failure. In some cases, failure is in tension through the bolt threads; in other instances, the threads of the nut and/or bolt strip. A tensile failure of the bolt is easily detected; however, a stripping failure develops with imperceptible reduction in torque and is difficult to identify, since some tension remains in the bolt. Therefore, when failure by over-tightening occurs or is imminent, a tensile failure of the bolt is preferable. To provide for this, nuts are specified to have a somewhat higher proof load than the bolts with which they are to be used.

As a nut is tightened against the resistance of the gripped material, the bolt lengthens within the grip. If the gripped material and the threads were completely rigid, one turn of the nut would cause the bolt to elongate one pitch. This does not happen, because some thread deformations occur in the bolt and nut. This diminishes the theoretical bolt elongation in the threaded portion.

Since the deformations of the threads are directly affected by the hardness of the nut or the bolt and the number of threads within the depth of the nut, calibration tests were performed on A325 high-strength bolts with minimum and maximum strength levels, and assembled with hex nuts and the thicker heavy hex nuts having various hardness values.[4.20] These tests showed that, with increasing nut hardness, the stripping strength of the connection also increases until the mode of failure changes to a tensile failure in the bolt thread. Also the bolt tension at one-half turn from a snug-tight condition increased with an increase in nut hardness, and higher bolt loads were observed in assemblies using high hardness bolts. For all bolt and nut combinations used in this study, the average bolt tension at snug-tight plus one-half turn was considerably above the required minimum tension.

On the basis of these tests[4.21] as well as other information the 1972 RCRBSJ specification[1.4] requires the use of heavy hex nuts. The latter have the advantage that they have the same across flat dimensions as the head of a heavy structural bolt which permits the use of a single wrench size for either bolt or nut.

4.10 BASIS FOR DESIGN RECOMMENDATIONS

The behavior of individual fasteners subjected to different types of loading forms a basis for developing design recommendations. This section summarizes the individual fastener strengths which are used in subsequent chapters to develop design recommendations.

4.10.1 Bolts Subjected to Tension

The tensile capacity of a fastener is equal to the product of the stress area A_s and its tensile strength σ_u. For design however, it is convenient to specify allowable tensile loads and stresses on the basis of the nominal bolt area A_b rather than on the stress area A_s. Such a transformation is readily performed, because the ratio of the stress area to the nominal bolt area varies from 0.75 for ¾-in.-diameter bolts to 0.79 for 1⅛-in.-diameter bolts. The maximum tensile load B_u of a fastener is given as

$$B_u = A_s \sigma_u \tag{4.2}$$

Expressed in terms of the nominal bolt area it yields a lower bound of

$$B_u = 0.75 A_b \sigma_u \tag{4.3}$$

For most bolt diameters, Eq. 4.3 yields a slightly conservative estimate of the tensile capacity of a bolt.

4.10.2 Bolts Subjected to Shear

The tension-type shear test was observed to provide a lower bound shear strength. The shear strength (in kilopounds per square inch) of a fastener was found to be independent of the bolt grade and equal to 62% of the tensile strength of the bolt material; hence

$$\tau_u = 0.62 \sigma_u \tag{4.4}$$

The shear resistance of a bolt is directly proportional to the available shear area and the number of critical shear planes. If a total of m critical shear planes pass through the bolt shank, the maximum shear resistance S_u of the bolt is equal to

$$S_u = m A_b (0.62) \sigma_u \tag{4.5}$$

When critical shear planes pass through the threaded portion of the bolt, the shear area is equal to the root area of the bolt, which is about 75 to 80% of the nominal bolt area. A lower bound to the maximum shear capacity of the bolt can be expressed as

$$S_u = (0.75) m A_b (0.62) \sigma_u \tag{4.6}$$

or

$$S_u = 0.47 m A_b \sigma_u \tag{4.7}$$

If the critical shear planes pass through the bolt shank as well as through the thread area, the total shear area is equal to the algebraic sum of the constituent shear planes.

4.10.3 Bolts Subjected to Combined Tension and Shear

An elliptical interaction curve was found to represent adequately the behavior of high-strength bolts under combined tension and shear. The equation was given in Section 4.2 as

$$\frac{x^2}{(0.62)^2} + y^2 = 1.0 \qquad (4.8)$$

where x is the ratio of the shear stress on the shear plane to the tensile strength and y represents the ratio of the tensile stress to the tensile strength (both computed on the stress area). Equation 4.8 relates the shear stress component to the critical tensile stress component. The product of ultimate stress and the appropriate area yields the critical shear and tensile load components.

References

4.1 R. J. Christopher, G. L. Kulak, and J. W. Fisher, "Calibration of Alloy Steel Bolts," *Journal of the Structural Division, ASCE*, Vol. 92, ST2, April 1966.

4.2 J. L. Rumpf and J. W. Fisher, "Calibration of A325 Bolts," *Journal of the Structural Division, ASCE*, Vol. 89, ST6, December 1963.

4.3 G. H. Sterling, E. W. J. Troup, E. Chesson, Jr., and J. W. Fisher, "Calibration Tests of A490 High-Strength Bolts," *Journal of the Structural Division, ASCE*, Vol. 91, ST5, October 1965.

4.4 J. J. Wallaert and J. W. Fisher, "Shear Strength of High-Strength Bolts," *Journal of the Structural Division, ASCE*, Vol. 91, ST3, June 1965.

4.5 R. T. Foreman and J. L. Rumpf, "Static Tension Tests of Compact Bolted Joints," *Transactions ASCE*, Vol. 126, Part 2, 1961, pp. 228–254.

4.6 R. A. Bendigo, R. M. Hansen, and J. L. Rumpf, "Long Bolted Joints," *Journal of the Structural Division*, Vol. 89, ST6, December 1963.

4.7 J. W. Fisher, P. Ramseier, and L. S. Beedle, "Strength of A440 Steel Joints Fastened with A325 Bolts," *Publications, IABSE*, Vol. 23, 1963.

4.8 E. Chesson, Jr., N. L. Faustino, and W. H. Munse, "High-Strength Bolts Subjected to Tension and Shear," *Journal of the Structural Division, ASCE*, Vol. 91, ST5, October 1965.

4.9 E. Chesson, Jr., and W. H. Munse, *Studies of the Behavior of High-Strength Bolts and Bolted Joints*, Engineering Experiment Bulletin 469, University of Illinois, Urbana, 1965.

4.10 J. Tajima, "*Effect of Relaxation and Creep on the Slip Load of High Strength Bolted Joints*," Structural Design Office, Japanese National Railways, Tokyo, June 1964.

4.11 P. C. Birkemoe and D. C. Herrschaft, "Bolted Galvanized Bridges—Engineering Acceptance Near," *Civil Engineering*, April 1970.

4.12 G. A. Maney, "Predicting Bolt Tension," *Fasteners*, Vol. 3, No. 5, 1946.
4.13 A. Pauw, and L. L. Howard, "Tension Control for High-Strength Structural Bolts," *Proceedings, American Institute of Steel Construction*, April 1955.
4.14 AREA Committee on Iron and Steel Structures, "Use of High-Strength Structural Bolts in Steel Railway Bridges," *American Railway Engineering Association*, Vol. 56, 1955.
4.15 F. P. Drew, "Tightening High-Strength Bolts," Proceeding Paper 786, ASCE, Vol. 81, August 1955.
4.16 E. F. Ball and J. J. Higgins, "Installation and Tightening of High-Strength Bolts," *Transactions, ASCE*, Vol. 126, Part 2, 1961.
4.17 M. H. Frincke, "Turn-of-Nut Method for Tensioning Bolts," *Civil Engineering*, Vol. 28, No. 1, January 1958.
4.18 G. C. Brookhart, I. H. Siddiqi, and D. D. Vasarhelyi, "Surface Treatment of High-Strength Bolted Joints," *Journal of the Structural Division, ASCE*, Vol. 94, ST3, March 1968.
4.19 W. H. Munse, "Structural Behavior of Hot Galvanized Bolted Connections," *Proceedings 8th International Conference on Hot Dip Galvanizing*, London, June 1967.
4.20 E. Chesson, Jr., W. H. Munse, R. L. Dineen, and J. G. Viner, "Performance of Nuts on High-Strength Bolts," *Fasteners*, Vol. 21, No. 3, 1967.
4.21 C. F. Krickenberger, Jr., E. Chesson, Jr., and W. H. Munse, *Evaluation of Nuts for Use with High-Strength Bolts*, Structural Research Series, No. 128, University of Illinois, Urbana, January 1957.
4.22 J. N. Macadam, *Research on Bolt Failures in Wolf-Creek Structural Plate Pipe*, Research Center Armco Steel Corporation, Middletown, Ohio, 1966.
4.23 W. K. Boyd and W. S. Hyler, "Factors Affecting Environmental Performance of High-Strength Bolts", *Journal of the Structural Division, ASCE*, Vol. 99, ST7, July 1973.
4.24 Subcommittee on Bolt Strength, "Delayed Fracture of High-Strength Bolts," *Society of Steel Construction of Japan*, Vol. 6, No. 52, Tokyo, June, 1970.
4.25 J. J. Wallaert and J. W. Fisher, "What Happens to Bolt Tension in Large Joints," *Fasteners*, Vol. 20, No. 3, 1965.
4.26 R. N. Allan and J. W. Fisher, "Bolted Joints with Oversize and Slotted Holes," *Journal of the Structural Division, ASCE*, Vol. 94, ST9, September 1968.
4.27 W. H. Munse, *The Case for Bolted Galvanized Bridges*, American Hot Dip Galvanizers Association, Inc., Washington, D. C., May 1971.
4.28 J. H. A. Struik, A. O. Oyeledun, and J. W. Fisher, "Bolt Tension Control with a Direct Tension Indicator," *Engineering Journal, AISC*, Vol. 10, No. 1, 1973.
4.29 W. S. Hyler, K. D. Humphrey, and N. S. Croth, *An Evaluation of the High Tensile Huck-Bolt Fastener for Structural Applications*, Report 72, Huck Manufacturing Co., Detroit, Michigan, March 1961.

Chapter Five

Symmetric Butt Splices

5.1 JOINT BEHAVIOR BEFORE SLIP

5.1.1 Introduction

A slip-resistant joint (also called friction-type joint) is one which has a low probability of slip at any time during the life of the structure. It is used where any occurrence of major slip would endanger the serviceability of the structure and therefore has to be avoided.

In a slip-resistant joint the external applied load usually acts in a plane perpendicular to the bolt axis. The load is completely transmitted by frictional forces acting on the contact area of the plates* fastened by the bolts. This frictional resistance is dependent on the bolt preload and slip resistance of the faying surfaces. The maximum capacity is assumed to have been reached when the frictional resistance is exceeded and overall slip of the joint occurs that brings the plates into bearing against the bolts.

Slip-resistant joints are often used in connections subjected to stress reversals, severe stress fluctuations, or other situations where slippage cannot be tolerated.

In the following sections, the different factors influencing the slip load of a connection are discussed.

5.1.2 Basic Slip Resistance

The slip load of a simple tension splice as shown in Fig. 5.1 is given by

$$P_{\text{slip}} = k_s m \sum_{i=1}^{n} T_i \qquad (5.1)$$

where

k_s = slip coefficient

m = number of slip planes

$\sum_{i=1}^{n} T_i$ = sum of the bolt tensions

* These include any connected part such as angles and other sections.

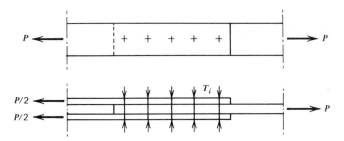

Fig. 5.1. Symmetric shear splice.

By assuming equal bolt tension in all bolts this reduces to

$$P_{\text{slip}} = k_s mn T_i \tag{5.2}$$

where n represents the number of bolts in the joint.

Equation 5.2 shows clearly that for a given number of slip planes and bolts, the slip load of the joint depends on the slip coefficient and bolt clamping force. For a given geometry, the slip load of the connection is proportional to the product of the slip coefficient k_s and bolt tension T_i.

Both the slip coefficient and bolt tension may vary from joint to joint. In general, a minimum required bolt tension is prescribed. However, it is well known that the actual bolt tension may exceed this value significantly. The slip coefficient is also variable from joint to joint and is highly dependent on factors such as the surface condition and treatment. Both quantities (k_s and T_i) show considerable variance. Consideration must be given to this variation when developing criteria for joint design.

5.1.3 Evaluation of Slip Characteristics

The slip coefficient k_s corresponding to the surface condition can only be determined experimentally. Usually a slip test is performed in which a symmetric butt joint is loaded in tension until slip of the connection occurs. The bolt preload, induced by the tightening process, is determined before the test is started. Once the slip load of the connection is known, the slip coefficient can be evaluated from Eq. 5.2.

$$k_s = \frac{P_{\text{slip}}}{mn T_i} \tag{5.3}$$

Most of the work done to determine the slip coefficient has been on symmetric butt joints of the type shown in Fig. 5.2. Both a two bolt specimen, type A, and a four bolt specimen, type B, have been used. The two

5.1 Joint Behavior Before Slip

standard test specimens with dimensions given in Fig. 5.2 are recommended for use with A325 as well as A490 bolts. Nearly identical specimens have been recommended in Europe by the European Convention of Steel Construction Federations.[5.30]

In fabricating and preparing the test specimens, care should be taken to ensure that the material and surface conditions of the test joints are representative of the conditions that occur in the field.

It is apparent from Eq. 5.3 that the value of the bolt clamping force T_i is of prime importance when determining the value of the slip coefficient k_s. Since the early stages of high-strength bolting, much attention has been directed to determine the axial force in a bolt installed in a joint. Up to the time of publication, no precise method is available. The best available method is to calibrate the bolts used in the test specimens.[4.2] This requires each bolt to be calibrated prior to installation in the test joint. The bolt clamping force should be within the elastic range if an accurate evaluation is made. Consequently, the bolts can be used more than once as long as the

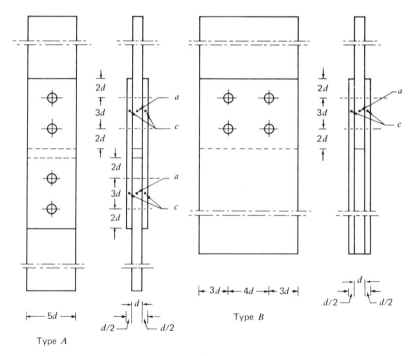

Fig. 5.2. Test specimens for determining the slip coefficient. Bolt diameter, d; hole diameter, $d + 1/16$ in.

grip length is not altered. If the bolts are tightened beyond the elastic limit load, permanent plastic bolt deformations will occur. In such cases an average bolt load-elongation curve for the lot to be used in the test joints has to be determined from a representative sample of bolts. The elongations of the bolts in the test joint can be related to the clamping force through this average bolt calibration curve. Because of inelastic deformations, the bolts can only be used once.

The calibration method requires special instrumentation and careful preparation of the bolts.[4.2] This can be avoided if bolts are used that indicate the bolt force. Several of these special bolt systems are presently available, for example, the TELL-TORQ fastener and the load indicating bolt. The TELL-TORQ fastener has an optical indicator in the bolt head which changes color depending upon the bolt load. If tightened only in the elastic range, the bolt can be used several times.

The load indicating bolt is a square headed bolt with notches at the corners of the bottom face of the bolt head. If the axial force exceeds a certain load, these notches are stressed beyond their elastic limit and visible deformations of the notches occur, indicating that a certain load level has been reached. Because the notches deform plastically, these bolts can be used once only.

Both load indicating bolt systems should be checked to assess the accuracy of the bolt in indicating the axial bolt force.

In a short-term static test the test specimens are subjected to gradually or incrementally increasing tensile loads. The displacements between points a and c (see Fig. 5.2) should be recorded at selected intervals of loadings.

In most slip tests on specimens without a protective coating on the slip surfaces, a sudden slip occurs when the slip resistance of the connection is exceeded. Coated specimens often do not exhibit sudden slip; the slipping builds up continuously as evidenced by cumulative microslips. In these situations, a slip load can only be arbitrarily defined. The load corresponding to a prescribed amount of slip, for instance, 0.005 in., is often used to define the slip load in these cases.

Other than major slip, creep of a connection might impair the serviceability of a joint as well. A creep test can be performed to evaluate the influence of sustained loading levels on the displacement of a joint. A constant load level is applied for a long period in a creep test and the observed displacements are evaluated.

5.1.4 Effect of Joint Geometry and Number of Faying Surfaces

The effects of joint geometry have been examined in numerous experimental studies. The significance of the influence of factors, such as number of

5.1 Joint Behavior Before Slip

bolts in a line and whether the bolts are arranged in compact patterns, has not been determined. An analysis of the slip coefficient in large bolted joints having clean mill scale surfaces yields an average slip coefficient 0.326 with a standard deviation of .069. For small joints these values were 0.340 and .071, respectively. In this analysis a large bolted joint was defined as having at least two lines of bolts parallel to the direction of the applied load with each line consisting of at least three bolts. Based on the results of this analysis, it was concluded that the number of bolts in a joint does not have a significant influence on the slip coefficient.

The slip resistance of a bolted joint is also proportional to the number of faying surfaces. Hence a multilap joint can resist slip with great efficiency. Tests have shown that the slip coefficient is not affected by the number of faying surfaces.[5.34]

5.1.5 Joint Stiffness

In slip-resistant joints the main plate and lap plates are compressed laterally by the initial clamping force. No relative displacement of the contact points on the surfaces takes place, and the joint may be considered equivalent to a solid piece of metal with a cross-section equal to the total area of the main and lap plates.

The stiffness of the joint, characterized by the slope of the load-deformation curve, decreases significantly if yielding in either the net or gross section occurs. This does not occur under working conditions, since the working load is much less than the yield load of the connection.

5.1.6 Effect of Type of Steel, Surface Preparation, and Treatment on the Slip Coefficient

One of the significant factors influencing the slip resistance of a connection is the slip coefficient k_s, as defined by Eq. 5.3. Because of its significant influence, much research has been done in the United States, Europe, and Japan to determine the magnitude of k_s for different steels, different surface treatments, and surface conditions.[4.5-4.7, 4.26, 5.1-5.17] The results of these studies have been used to evaluate the slip coefficient for a number of surface conditions.

It is clear that to determine a reliable value of the slip coefficient k_s, an accurate estimate of the initial clamping force must be known. Therefore, only tests where the actual clamping force in the bolts was measured were considered in the following analysis. Data obtained from tests in which bolts were installed using torque control were not considered.

In many cases structural members are bolted together without special treatment of the faying surfaces. A natural faying surface is provided by

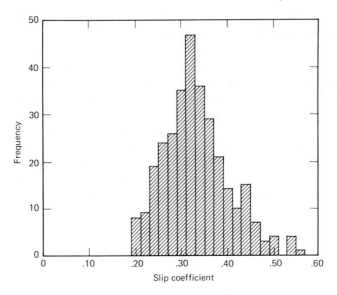

Fig. 5.3. Distribution of slip coefficient for clean mill scale surfaces. Clean mill scale surfaces: A7, A36, A440, Fe37, and Fe52 steel. Number of tests, 312; average, 0.336; standard deviation, 0.070.

clean mill scale. Only the loose mill scale and dirt is removed by hand wire brushing. Grease originating from the fabrication process is removed with a solvent. An analysis of the available data shows that the clean mill scale condition for A7, A36, and A440 structural steels yield an average slip coefficient k_s of 0.325 with a standard deviation of 0.06. Tests performed in Europe on Fe37 and Fe52 steels comparable to A7, A36, and A440 steels exhibited similar results. If all the available data on A7, A36, Fe37, A440, and Fe52 steel are considered, an average value of k_s equal to .336 is obtained with a standard deviation of 0.070. Figure 5.3 shows the frequency distribution of the slip coefficient as derived from the 312 tests.

If the mill scale is removed by brushing with a power tool, a shiny clean surface is formed that decreases the slip resistance. Joints tested at Lehigh University with such semi-polished faying surfaces indicated a decrease in friction resistance of 25 to 30% as compared with normal hand brushed mill scale surfaces.[5,6] This decrease is mainly due to the polishing effect of the power tool; the surface irregularities, which are essential for providing the frictional resistance are reduced, causing a decrease in k_s.

5.1 Joint Behavior Before Slip

Many tests have shown that blast cleaning greatly increases the slip resistance of most steels as compared with the clean mill scale condition.[5.5, 5.11] An analysis of available data yielded an average value k_s equal to 0.49 for A7, A36, and Fe37 steels with blast cleaned surfaces. The frequency distribution of the test results is shown in Fig. 5.4. It is apparent that the frequency distribution is somewhat skewed. This is reasonable, since the higher values could be influenced by yielding of the steel. The friction coefficient for blast-cleaned A440 and Fe52 steel should not differ from the value reported for blast cleaned A7, A36, and Fe37 steel surfaces.

The magnitude of k_s for shot-blasted surfaces is greatly affected by the type and condition of grit or material that is employed to clean the surface. The condition of the cleaning material determines whether the surfaces are polished or left with a rough texture that is more slip resistant.

Tests on A514 constructional alloy steel showed an average slip coefficient of 0.33 for steel grit-blasted surfaces. Although not much experimental evidence is available at the present time, these results show that grit blasting of quenched and tempered alloy steel as compared with lower strength steel has less effect on the slip coefficient. This indicates that the hardness of the surface influences the roughness achieved by the blast cleaning.

In most field situations, structural members are exposed to the atmosphere for a period of time before erection. During this period unprotected blast-cleaned surfaces are highly susceptible to surface corrosion. To simulate this field condition, tests were performed in which the blast-cleaned surfaces were stored in the open air for different periods, before being assembled and tested.[5.11, 5.15] These test specimens were bolted up without

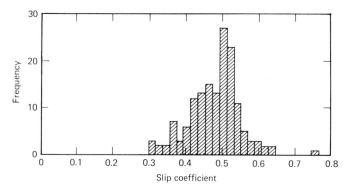

Fig. 5.4. Distribution of slip coefficient for blast-cleaned surfaces. Blast-cleaned surfaces: A7-A36-Fe37 steel; number of tests, 168; average, 0.493; standard deviation, 0.074.

Table 5.1. Summary of Slip Coefficients[a]

Type Steel	Treatment	Average	Standard Deviation	Number of Tests
A7, A36, A440	Clean mill scale	0.322	.062	180
A7, A36, A440	Clean mill scale	0.336	.070	312
Fe37, Fe52	Red lead paint	0.065	—	6
A7, A36, Fe37	Grit blasted	0.493	.074	168
A7, A36, Fe37	Grit blasted, exposed (short period)	0.527	.056	51
A514	Grit blasted	0.331	.043	19
A7, A36	Semi polished	0.279	.043	12
A7, A36, Fe37	Hot dip galvanized	0.184	.041	27
	Vinyl treated	0.275	.023	15
	Cold zinc painted	0.30	—	3
	Metallized	0.48	—	2
	Rust preventing paint	0.60	—	3
	Galvanized and sand blasted	0.34	—	1
	Sand blasted and treated with linseed oil (exposed)	0.26	—	3
	Sand blasted	0.47	—	3

[a] Determined from tension type specimens.

wire brushing or otherwise disturbing the rusted surfaces. The results indicated that with increased exposure time the slip coefficient of the blast cleaned surfaces was reduced considerably. A value that is about the same as for clean mill scale condition resulted. Removing the rust by wire brushing improved the slip resistance.

If rust forming on the blast-cleaned faying surfaces cannot be tolerated, a protective coating can be applied to the surfaces. These protective treatments alter the slip characteristics of bolted joints to varying degrees. Tests have been performed to evaluate the behavior of bolted joints in which the faying surfaces were galvanized, cold zinc painted, metallized, treated with vinyl wash or linseed oil, or treated with rust preventing paint.[5.5, 5.12, 5.15, 5.17, 5.21, 5.37, 5.46] The results of these tests are summarized in Table 5.1. Some of the values listed in this summary were determined from a rather small number of tests. They provide only an indication of the

5.1 Joint Behavior Before Slip

magnitude of the slip coefficient. Chapter 12 describes in greater detail the influence of surface coatings on the slip resistance of bolted joints.

5.1.7 Effect of Variation in Bolt Clamping Force

Besides the slip coefficient k_s, the initial bolt clamping force T_i is one of the major factors governing the slip load of a connection as is apparent from Eq. 5.2. A variation in the initial clamping force directly affects the slip load of the connection. Experience has shown that the actual bolt tensions in a joint usually exceed the minimum tension required by specifications. This results from different tightening methods and variations in the mechanical properties of the bolts.

Bolts can be tightened by either the turn-of-nut method or with calibrated wrenches. The turn of nut is primarily based on an elongation control, whereas the calibrated wrench method is based on controlling the applied torque. The two methods do not necessarily yield the same bolt tension as illustrated in Fig 5.5. Here the influence of the tightening method on the achieved bolt tension is shown for two bolt lots having different mechanical properties. When the calibrated wrench method is used the bolt tension T_{iC} is about the same for both lots since the wrench is

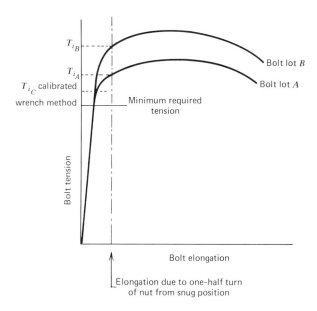

Fig. 5.5. Influence of tightening method on the achieved bolt tension for different bolt lots.

adjusted for each lot. However, if the turn-of-nut method is employed the average elongation of the bolts will be about the same for both lots. Consequently the bolt tensions T_{iA} and T_{iB} will differ as illustrated in Fig. 5.5.

i. Turn-of-the-Nut Method. Figure 5.5 illustrates that the tensile strength of the bolt is a significant factor influencing the induced bolt tension when the turn-of-nut method is used. An increase in tensile strength leads to an increase in initial bolt tension in an installed bolt. An analysis of the data obtained from several bolt lots used in joints and calibration tests at Lehigh University indicates that the relationship between the tensile strength and initial bolt tension can be approximated by the straight line relationship given in Fig. 5.6. The tensile strength of a bolt was determined from static tension tests on representative samples. The induced bolt tension at one-half turn from the snug position can be derived from the measured average tensile force in bolts installed in joints or by torquing the bolts in a hydraulic calibrator. The data plotted in Fig. 5.6 show clearly that torquing a bolt one-half turn from the snug position in a gripped

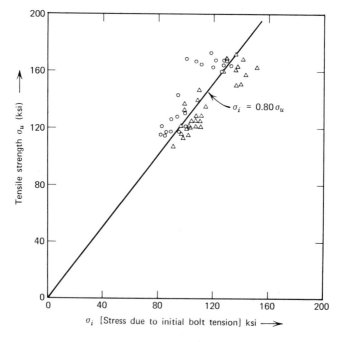

Fig. 5.6. T_i versus T_u in Lehigh tests. ○ Data from calibration tests; △ data from test joints.

5.1 Joint Behavior Before Slip

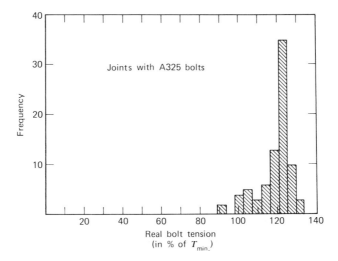

Fig. 5.7. Distribution of initial bolt force in test joints with A325 bolts installed by turn of nut. Number of tests, 81; average value, 120.2%; standard deviation, 9.1%.

material such as a joint leads to a higher tension stress than obtained by torquing the bolt one-half turn in a hydraulic calibrator. This is mainly due to the difference in stiffness of the gripped material.[4.1] Based on a least square fit of all the data plotted in Fig. 5.6, the relationship between σ_i and σ_u was determined as

$$\sigma_i = 0.80\sigma_u \tag{5.4}$$

Most of the data obtained from calibration tests in a hydraulic calibrator yield smaller bolt tensions compared with the data obtained from test joints (see Fig. 5.6). Hence including the above, data tend to yield a conservative estimate of the average bolt tension in a joint based on the average tensile strength of the bolts.

The actual bolt tension using the turn-of-nut method may exceed substantially the required minimum tension. This is illustrated in Fig. 5.7 where test data obtained from joints assembled with A325 bolts are shown. The bolt tension on the horizontal axis is plotted as a percentage of the minimum required bolt tension. The average bolt tension in these joints was about 20% greater than the required minimum tension. In joints assembled with A490 bolts by the turn-of-nut method an average bolt tension of 26% greater than the required minimum tension was observed. The bolts used in these tests were purposely ordered to minimum strength

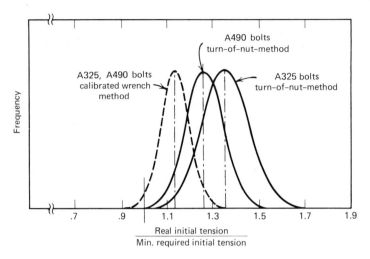

Fig. 5.8. Distribution curves $T_i/T_{i\,\text{spec}}$ for different installation procedures.

requirements of the applicable ASTM specification. Although the actual tensile strength of the bolts exceeded the required tensile strength (3% for A325 and 10% for the A490 bolts), it was less than the average tensile strength of production bolts.

Since the average tensile strength of A325 bolts is

$$\sigma_{u\ \text{real}} = 1.183 \sigma_{u\ \text{specified}}$$

and the average clamping force is about 80% of the actual tensile strength, it follows that the installed bolt tension σ_i is about equal to 0.95 $\sigma_{u\,\text{specified}}$. Present specifications require the minimum bolt tension to equal or exceed 70% of the specified tensile strength. Hence the average actual bolt tension will likely exceed the required minimum bolt tension by approximately 35% when the turn-of-nut method is used to install the bolts.

A similar analysis of A490 bolts shows that the average initial bolt tension can be expected to exceed the minimum required bolt tension by approximately 26%.

To characterize the frequency distribution of the ratio $T_i/T_{i\,\text{specified}}$, the standard deviation as well as the average value of the ratio are required. These have been estimated for both A325 and A490 bolts from test results. Data obtained at the University of Illinois and Lehigh University showed that the standard deviation of the ratio $T_i/T_{i\,\text{specified}}$ from average values was 8 and 10% for A325 and A490 bolts, respectively. By assuming a normal distribution, the frequency distribution curve of the ratio $T_i/$

5.1 Joint Behavior Before Slip

$T_{i \text{ specified}}$ can be defined. Figure 5.8 shows these curves for A325 and A490 bolts. The figure shows that bolts installed by the turn-of-nut method will provide a bolt tension which exceeds the minimum required tension.

It was noted earlier that the average tensile strength of production A325 bolts exceeds the required tensile strength by approximately 18%. This was observed for bolt sizes up to 1-in. diameter. For A325 bolts greater than 1 in. the range of actual over specified minimum ultimate strength is even more favorable. The extra strength of bolts larger than 1 in. was not considered.

   ii. Calibrated Wrench Method. A variation in mechanical properties of bolts does not affect the average installed bolt tension when the calibrated wrench is used. However, since this method is essentially one of torque control, factors such as friction between the nut and the bolt and between the nut and washer are of major importance. An analysis of 231 tests, in which single bolts were subjected to a constant predetermined applied torque, showed that the standard deviation of the recorded bolt tension equaled 9.4% of the recorded value.[4.13, 5.35, 5.36] It was observed that the variation of the average clamping force for a joint decreases, depending on the number of bolts in the joint. For a joint having five bolts, the standard variation of the average bolt clamping force becomes 5.6% of the required mean value.

Since variations in bolt tension do occur as a result of variations in thread mating, lubrication, and presence or absence of dirt particles in the threads, the wrench is required to be adjusted to stall at bolt tensions which are 5 to 10% greater than the required preload.

Tests have indicated that installing a bolt in a joint leads to a higher bolt tension as compared with torquing the bolt in a hydraulic calibrator. This difference is about equal to 5.5%. Consequently the average clamping force in a five-bolt joint, with bolts installed by the calibrated wrench with a setting 7.5% greater than the required preload is equal to

$$(0.7\sigma_u)(107.5)(1.055) = 0.796\sigma_u$$

or $1.13\sigma_{\text{spec. min.}}$.

The standard deviation is equal to about 6%. The corresponding frequency distribution curves of the ratio $T_i/T_{i \text{ specified}}$ are also shown in Fig. 5.8.

5.1.8 Effect of Grip Length

Grip length does not have a noticeable influence on the behavior of friction-type joints. The only point of concern is the attainment of the desired clamping force. When the bolt length in the grip is greater than eight times the diameter, one-half turn from the snug position may not provide the required preload. The greater bolt length requires an increased amount of deformation. To provide this increased bolt elongation an additional increment of nut rotation is required. When tightening is done by turn-of-nut method, present specifications require a two-thirds nut rotation instead of one-half to achieve at least minimum bolt tension in joints with large grips.

5.2 JOINT BEHAVIOR AFTER MAJOR SLIP

5.2.1 Introduction

When the frictional resistance of a joint is exceeded, a major slip occurs between the connected elements. Movement is stopped when the hole clearance is taken up and the bolts are in bearing. From this stage on the load is mainly transferred by means of shear and bearing. This has led to the concept of a "bearing-type" joint. In bearing-type joints the shear strength of the fasteners is the critical parameter, not the bolt preload.

High-strength bolts are tightened to at least the minimum required tension in all bolted joints, since this increases the overall joint rigidity, prevents confusion during installation, and leads to a better stress pattern throughout the joint under working load conditions. Furthermore, it provides security against nut loosening in situations where the joint is repeatedly loaded or subjected to vibration. The axial preload in the bolt has no noticeable effect on the ultimate shear strength however.[4.4]

5.2.2 Behavior of Joints

The applied load in bearing type joints may be transferred either by friction or by shear and bearing, depending on the magnitude of the load and the faying surface condition. In most joints a combination of both effects is likely to occur under normal service loads.

Initially the load is transferred by friction forces at the ends of the joints. This is known from both elastic studies as well as experimental investigations.[5.5] As the load is increased, the zone of friction extends toward the center of the joint as illustrated in Fig. 5.9. Eventually the maximum frictional resistance is exceeded at the ends, and small displacements of contact points on the faying surfaces takes place. This is illustrated schematically in Fig. 5.9 as case 2. As load on the connection is increased, the slip zone proceeds inward from the ends toward the center of the joint. When the applied load exceeds the frictional resistance over the entire faying surface of the connection, large relative displacements occur. This movement, called major slip, may be equal to the full hole clearance but usually it is considerably smaller.[4.6-4.7, 5.6, 5.12, 5.25, 5.45]

5.2 Joint Behavior After Major Slip

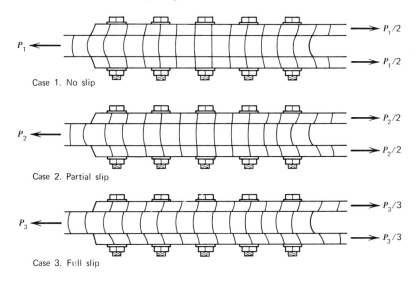

Schematic representation of three displacement conditions of a high-strength bolted joint.

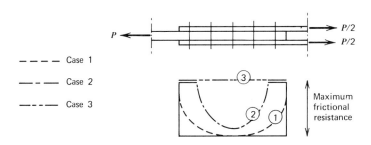

Fig. 5.9. Distribution of friction forces for cases 1 to 3.

When major slip occurs, only the end bolts may come into bearing against the main and splice plates. As the applied load is increased, the end bolts and holes deform further until the succeeding bolts come into bearing. This process continues until all of the bolts are in bearing as illustrated for case 3 in Fig. 5.9.

Further application of load causes each bolt to deform in proportion to the force it transfers. The deformation of a bolt during this stage depends on the difference in the pitch elongations between any two adjacent transverse rows. Since these deformations are greater at the ends of the joint, the end bolts are carrying greater loads. A leveling out occurs if the bolts are ductile in shear, as is illustrated in Fig. 5.10a. Eventually the end pitches have such large displacement and differential elongation that the end bolts fail in shear.

In short connections, with only a few fasteners in line, almost complete equalization of load is likely to take place before bolt failure occurs. Failure in this case appears as a simultaneous shearing of all the bolts.

In longer joints, the end fasteners may reach their critical shear deformation and fail before the full strength of each fastener can be achieved. The large shearing deformations of the end bolts and the greater elongation of the end holes is shown in Fig. 5.10b. The remaining bolts are usually not capable of taking much additional load without incurring failure themselves in a sequential fashion. The sequential failure of fasteners in long connections is called "unbuttoning." This phenomenon is predicted by theoretical analysis and has been witnessed in tests of long bolted and riveted joints.[4.6, 4.7, 5.6, 5.12, 5.21, 5.25]

Figure 5.11 shows load-deformation curves for two A7 steel ($\sigma_y = 36$ ksi) joints connected with A325 bolts. Figure 5.11a is the test curve of a joint with semi-polished faying surfaces. A gradual slip occurred as load was applied. The second joint had clean mill scale surfaces and exhibited a sudden slip as illustrated in Fig. 5.11b.

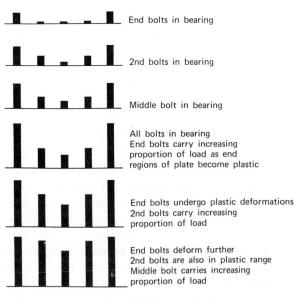

(a)

5.2 Joint Behavior After Major Slip

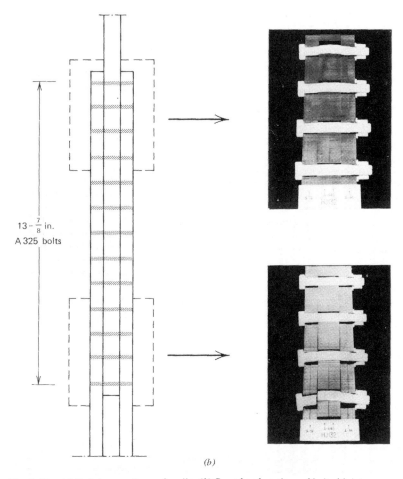

Fig. 5.10. (a) Bolt forces after major slip. (b) Sawed end sections of bolted joint.

High-strength bolts are usually placed in holes that are nominally $\frac{1}{16}$ in. larger than the bolt diameter. Therefore, the maximum slip that can occur in a joint is equal to $\frac{1}{8}$ in. However, field practice has shown that joint movements are rarely as large as $\frac{1}{8}$ in. and average less than $\frac{1}{32}$ in.[5.45] In many situations the joint will not slip at all under live loads because the joint is often in bearing by the time the bolts are tightened. This might be due to small misalignments inherent to the fabrication process. In addition, under the dead load, slip may have occurred before the bolts in the joint

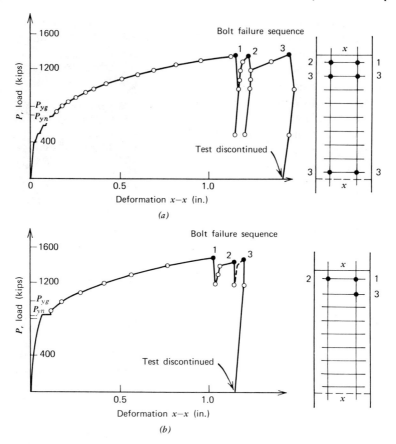

Fig. 5.11. Typical load-deformation curves for different surface conditions. (*a*) Semi-polished surfaces; (*b*) clean mill scale surfaces.

were tightened. Generally, slips under live loads are so small that they seldom have a serious effect on the structure.

Tests have indicated that when bolts undergo large shearing deformations, clamping force is relaxed. By the time a bolt approaches its ultimate shearing load, practically no clamping force exists.[4.6, 4.7] In addition, extensive yielding of the net section may also reduce the clamping force because of the concurrent reduction in plate thickness. As a consequence, there is negligible frictional resistance in the vicinity of a bolt at ultimate load.

5.2.3 Joint Stiffness

The stiffness of a bearing-type joint is equal to the stiffness of similar slip resistant joints until slip occurs. Slip of the connection brings one or more

5.2 Joint Behavior After Major Slip

bolts into bearing and results in motion of the lap plates with respect to the main plates. The stiffness of the joint, characterized by the slope of the load elongation curve, is not affected by slip. This is illustrated in Fig. 5.12. Only yielding of the gross and net sections caused a significant change in the slope of the load elongation curve.

The load versus deformation curves shown in Figs. 5.11 and 5.12 show a distinct slip. In most situations the slips are so small that they have no significant effect on the structure. The joint stiffness of a bearing-type joint is about the same as the stiffness of a similar slip-resistant joint if the joint is erected in bearing.

5.2.4 Surface Preparation and Treatment

The level of slip resistance does not influence the ultimate strength of a shear splice (see Fig. 5.11). Consequently, the surface condition of the connected plates is not critical except for slip critical joints. Hence painted, galvanized, or other surface conditions that may result in a low slip coefficient do not influence the ultimate strength of bolted joints.

The slip resistance is an important factor influencing the joint behavior under repeated loadings. Depending on the ratio between the slip resistance and applied load, failure may occur through either the net or gross section area. A more detailed discussion on this is given in Section 5.3.

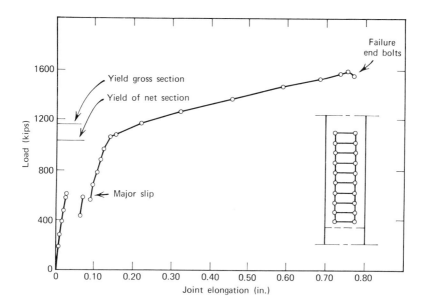

Fig. 5.12. Typical load-deformation curve of high-strength bolted joint.

5.2.5 Load Partition and Ultimate Strength

Theoretical studies of mechanically fastened joints have been made since the beginning of this century. A linear, elastic relationship between load and deformation was assumed in early studies. However, since the early 1960s, mathematical models that establish the relationships between deformation and load throughout the elastic and inelastic range for component parts of joints have been developed.[5.21] The method of analysis is summarized briefly in this section for a double shear symmetrical butt joint. For purposes of analysis the joint is divided into gage strips and it is assumed that all gage strips are identical in behavior. Test results have indicated that this is a reasonable approximation.

The theoretical solution of the load partition at ultimate load is based on the following major assumptions: (1) the fasteners transmit all the applied load by shear and bearing once major slip has occurred; and (2) the frictional forces may be neglected in the region for which the solution is intended, the region between major slip and ultimate load.

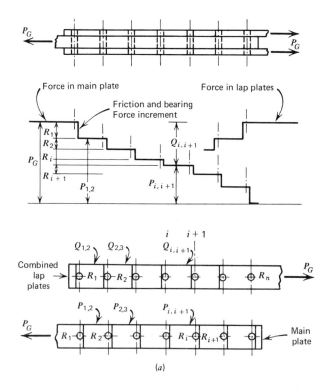

(a)

5.2 Joint Behavior After Major Slip

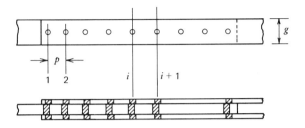

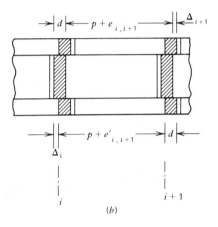

Fig. 5.13. Idealized load transfer diagrams and deformations in bolts and plates. (a) Load transfer. (b) Deformations in bolts and plates.

The solution is obtained by formulating the following two basic conditions: (1) satisfying the condition of equilibrium, and (2) assuring that continuity will be maintained throughout the joint length for all load levels. These conditions, coupled with initial value considerations such as the ultimate strength of the plate and the ultimate strength and deformation capacity of the critical fastener, yield the solution.

The equilibrium conditions can be visualized with the aid of Fig. 5.13a. The load per gage strip in the main plate between bolts i and $i+1$ is equal to the total load on this strip, P_G, minus the sum of the loads on all bolts, ΣR_i, preceding the part of the joint considered, that is, between i and $i+1$:

$$P_{i,i+1} = P_G - \sum_{i=1}^{i} R_i \qquad (5.5)$$

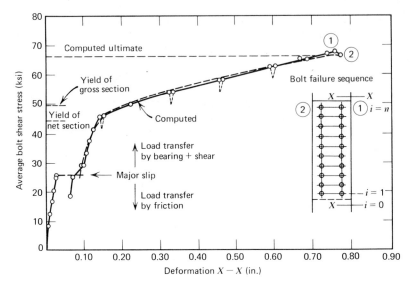

Fig. 5.14. Comparison of theoretical and experimental results.

The load per gage in the lap plates between bolts i and $i + 1$ is equal to the sum of the loads transmitted to the lap plate by all the bolts preceding the part of the joint considered. Hence

$$Q_{i,i+1} = \sum_{i=1}^{i} R_i \tag{5.6}$$

The compatibility equations can be formulated by considering the deformations illustrated in Fig. 5.13b. As a result of the applied load, the main plate will have elongated so that the distance between the main plate holes is $p + e_{i,i+1}$. The lap plate will have elongated and the distance between the lap plate holes is $p + e'_{i,i+1}$. The bolts will have undergone deformations Δ_i which include the effects of shear, bending, and bearing of the fastener and the localized effect of bearing on the plates. It is assumed that the deformations of the fastener Δ_i are the same whether considered at the hole edge (fastener surface) or the center line of the fastener. A further, detailed analysis of the parameters, included in Δ_i and Δ_{i+1} is given in Refs. 5.21 and 5.22.

The compatibility condition between points i and $i + 1$ yields

$$\Delta_i + e'_{i,i+1} = \Delta_{i+1} + e_{i,i+1} \tag{5.7}$$

5.2 Joint Behavior After Major Slip

If the plate elongations are expressed as functions of load in the segments of the joint between fasteners, and the fastener deformations as functions of the fastener loads, Eq. 5.7 can be written as

$$f(R_i) + \Psi(Q_{i,i+1}) = f(R_{i+1}) + \Phi(P_{i,i+1}) \tag{5.8}$$

in which $f(R_i)$ and $f(R_{i+1})$ represent the bolt deformations, $\Phi(P_{i,i+1})$, the main plate elongation and $\Psi(Q_{i,i+1})$ the lap plate elongation.

Equation 5.8 can be written for each section of the joint, giving $n - 1$ simultaneous equations. These, with the equation of equilibrium,

$$P_G - \sum_{i=1}^{n} R_i = 0 \tag{5.9}$$

may be solved to give the loads acting on the fastener when the relationship between the load and elongation for the various components are known.[5.21, 5.22] With this information, the total load acting on the joint may be found for a given deformation, and finally, the ultimate strength, the load at failure, may be determined.

The solution of the equilibrium and compatibility equations are lengthy and laborious, especially for long joints with many fasteners. Obviously

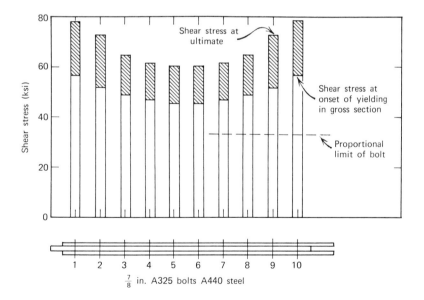

Fig. 5.15. Load partition in joint with 10 fasteners in a line.

such solutions are not practical for design purposes. However, the theoretical solution for the ultimate strength and load partition has been accomplished by computer studies and verified by comparing the theoretical results with the results of tests of large steel joints with yield strengths ranging from 33 to 100 ksi.[4.6, 5.6, 5.12] In all cases the theory and test results were in good agreement. Fig. 5.14 shows the experimental and the theoretical load deformation curve for a bolted joint with two lines of ten $\frac{7}{8}$-in. A325 bolts per line. The yield stress of the plate material was about 44 ksi and the ratio of the net section area to the gross section area, denoted as the A_n/A_g ratio, was equal to 1 to 1.10 for this particular joint. The theoretical loads carried by each fastener at two stages of loading are shown in Fig. 5.15. The end fasteners are obviously the critical ones.

5.2.6 Effect of Joint Geometry

By means of the theoretical solution summarized in Section 5.2.5 it is possible to study the effect of material and geometrical parameters that govern the joint behavior. In this article the significance of a number of parameters such as the joint length, the pitch, the relative proportions between the net tensile area of the plate, and the total bolt shear area (A_n/A_s ratio), the type of connected material, the A_n/A_g ratio, and the fastener pattern are examined briefly. A more detailed analysis of these parameters is presented in Refs. 5.21 and 5.23. All the hypothetical studies are based on minimum strength plate and fasteners and provide a lower bound to the joint strength.

i. Effect of Joint Length. Theoretical as well as experimental studies have shown that the joint length is an important parameter which influ-

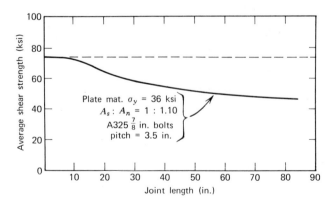

Fig. 5.16. Effect of joint length on ultimate strength.

5.2 Joint Behavior After Major Slip

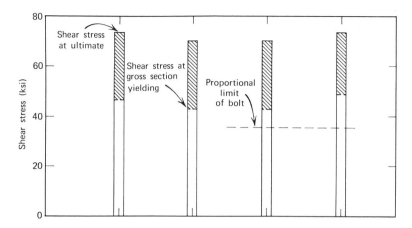

Fig. 5.17. Load partition in joint with four fasteners in line. Plate material $\sigma_y = 36$ ksi. Fastened by 7/8 in. A325 bolts.

ences the ultimate strength of the joint. Depending on factors such as type of plate material and fastener deformation capacity, a simultaneous shearing of all the bolts or a sequential failure (unbuttoning) of all the bolts may occur depending on the joint length.

For a given number of fasteners, the joint length is a function of the fastener spacing (pitch). A constant pitch of 3.5 in. and a ratio of bolt shear

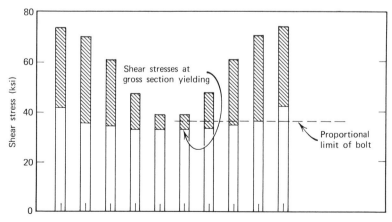

Fig. 5.18. Load partition in joint with 10 fasteners in line. Plate material $\sigma_y = 36$ ksi. Fastened by 7/8-in. A325 bolts.

area to net tensile area equal to 1.10 was used in theoretical studies to illustrate the effect of joint length. The joint material has a yield strength of 36 ksi and is fastened by $7/8$-in. A325 bolts. If the design stress of the plate material is taken as 24 ksi, then an A_n/A_s ratio of 1.10 yields an average shear stress of about 22 ksi for the fasteners.

The results of the theoretical studies are summarized in Fig. 5.16 where the average fastener shear at ultimate load is plotted as a function of the joint length. The longer joints show a decrease in average shear strength. Short or "compact" joints were affected to a negligible extent. Joints up to 10 in. in length provided the same average shear strength. As the number of fasteners was increased, Fig. 5.16 indicates that a decrease in the average strength occurred at a decreasing rate.

The reason for the decrease in shear strength with increased length of the joint is illustrated in Figs. 5.17, 5.18, and 5.19. The computed shear stress in each bolt at two different loading stages are shown for joints having four, ten, and 20 fasteners in a line. The two stages are (1) onset of yielding in the gross section of the plate designated by the end of the open portion of the bar, and (2) bolt stress at ultimate load (designated by the top of the shaded portion). Figure 5.17 shows that almost complete redistribution of bolt forces has taken place in the four-bolt joint, since all fasteners are carrying about the same load at ultimate. As joint length is increased, Figs. 5.18 and 5.19 show that the fasteners near the center of the joint carry only about half the forces carried by the end fasteners. Consequently, the average shear stress on the fastener is significantly reduced.

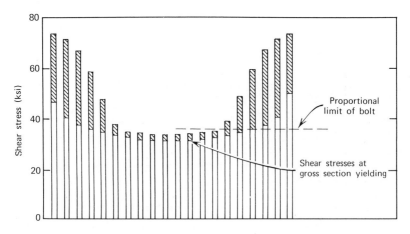

Fig. 5.19. Load partition in joint with 20 fasteners in line. Plate material σ_y = 36 ksi. Fastened by 7/8 in. A325 bolts.

5.2 Joint Behavior After Major Slip

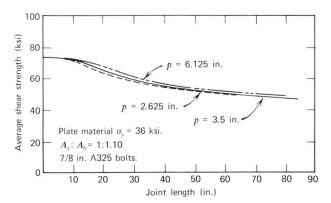

Fig. 5.20. Effect of pitch on the ultimate strength of A7 steel joints. Plate material σ_y = 36 ksi. $A_s:A_n$ = 1:1.10; 7/8-in. A325 bolts.

Theoretical investigations to determine the influence of joint length on the load distribution in joints of steel with a yield stress other than 36 ksi have been made.[5.23-5.25] Steels with yield stress ranging from 36 to 100 ksi, as well as hybrid steel joints were examined, and the results indicated a load distribution similar to the one described previously for a 36-ksi yield stress plate material.

ii. Effect of Pitch. The pitch is the distance between centers of adjacent fasteners along the line of principal stress. To determine the effect of the fastener pitch, analytical studies were made for joints with different fastener spacings, bolt grades, and connected material.[5.6, 5.21] The results of an analysis of a 36-ksi yield stress plate material, connected by 7/8-in. A325 bolts are summarized in Fig. 5.20. Three different fastener spacings equal to three, four, and seven times the bolt diameter were examined. The curves indicate that the change in shear strength with length is not greatly influenced by the pitch of the fasteners. If a joint with a given number of fasteners in a line is shortened by reducing the pitch between bolts, equal or greater strength results from the decrease in length. These studies have shown that pitch length, per se, is not an important variable. For a given A_n/A_s ratio, the shear strength is controlled by total joint length rather than by pitch length.

iii. Effect of Variation in Relative Proportions of Shear and Tensile Areas. There are two possible modes of failure in a bearing-type connection subjected to a tensile load. If the differential plate strains near the ends of a joint are high as compared to those in the central portion, the shear

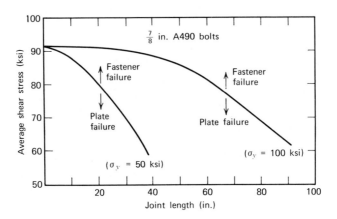

Fig. 5.21. Failure mode boundary (τ vs. L).

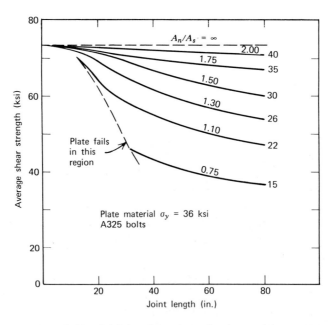

Fig. 5.22. Effect of variation of A_n/A_s ratio—structural carbon steel fastened by A325 bolts.

5.2 Joint Behavior After Major Slip

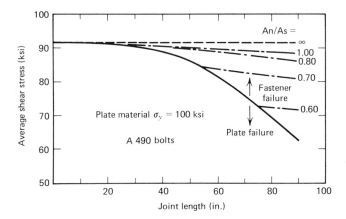

Fig. 5.23. Effect of variation in A_n/A_s ratio—quenched and tempered alloy steel fastened by A490 bolts.

failure of a single end fastener can occur. The resulting distribution of load from the failed connector to those remaining usually causes a sequential failure or unbuttoning, and little if any additional strength is available. When the tensile capacity of the plate at its net section is less than the shear capacity of the fasteners, failure will obviously occur by fracture of the plate.

Establishing the plate-failure-fastener-failure boundary line cannot be done directly, since joint length and the ratio of shear to net area both influence the shear strength. When the bolt shear strength and the plate capacity converge, a point on the boundary is determined. This process can be repeated for various joint lengths until the complete curve has been defined as shown in Fig. 5.21. For comparative purposes curves for steels with a yield stress of 50 and 100 ksi are shown.

It has been theoretically predicted and experimentally verified that as the A_n/A_s ratio for a joint is increased for any given joint length, the average shear strength also increases.[5.23] Figure 5.22 summarizes the results of analytical studies on joints of a plate material having a 36-ksi yield stress and fastened with A325 bolts. An increase in the A_n/A_s ratio corresponds to an increase in the net tensile area. The ideal case of equal load distribution among fasteners occurs when $A_n/A_s = \infty$. This represents a perfectly rigid joint. For any lesser value of A_n/A_s the fasteners carry unequal load depending on the joint length. Figure 5.23 shows the effect of a variation in the A_n/A_s ratio for joints fastened by A490 bolts. A yield stress of 100 ksi was assumed for the plate material.

Both Figs. 5.22 and 5.23 illustrate that with an increase in the net plate area, the average shear strength of the fasteners for the longer joints is greater. For shorter joints, plate failure may occur before bolt failure. Only an increase in joint length can cause bolt failure.

This examination has illustrated that it is not possible to maintain a uniform condition for both bolts and plates. When joints are short, the usual plate geometry will cause plate failure to occur. As joint length is increased, a balanced condition can occur for a specified length. For longer joints, bolt failure will be the governing mode. For design, the achievement of a proper balance between these failure conditions is required.

iv. **Effect of Variation in Gage Width and A_n/A_g Ratio.** In evaluating the performance of any structure, it is usually considered desirable for the system to have capacity for distortion or geometrical adjustment before failure by fracture. In an axially loaded structure, this means that the connections should permit yielding to occur in the gross cross-section of the member, before the joint fails through the net section if at all possible.[5.23] This requirement is satisfied if

$$\frac{A_n}{A_g} \geq \frac{\sigma_y}{\sigma_u} \qquad (5.10)$$

It is apparent that, depending on the type of steel, Eq. 5.10 leads to different minimum A_n/A_g ratios. Based on the minimum specified yield and tensile strengths for the types of steel, the A_n/A_g ratio has to equal or exceed 0.60 or 0.69 for structural carbon steel and high-strength steel, respectively, to achieve yielding of the gross section before failure of the net section occurs. For joints made of quenched and tempered alloy steel the minimum A_n/A_g ratio is equal to 0.87.

These A_n/A_g requirements are satisfied in most structures of carbon or high-strength steel. However, it has been shown that for A514 steel (yield stress 90 to 100 ksi) tension members, current practice commonly will lead to the situation, wherein the member will fail through the net section before yielding is reached in the gross section, unless special provisions such as upset ends or other changes in cross section are made to ensure yielding of the gross section before the net section fails.[5.23, 5.38] If yielding in the gross section cannot be achieved, a greater margin against ultimate is needed.

The A_n/A_g ratio depends on factors such as the gauge width of the joint and the hole diameter. For a constant hole diameter, an increase in the gauge width g increases the A_n/A_g ratio; therefore, gross section yielding is more likely to occur before failure of the net section. An increase in gauge width also tends to decrease slightly the tensile strength of the net section. However, this is not critical, since gross section yielding of the member can be expected.

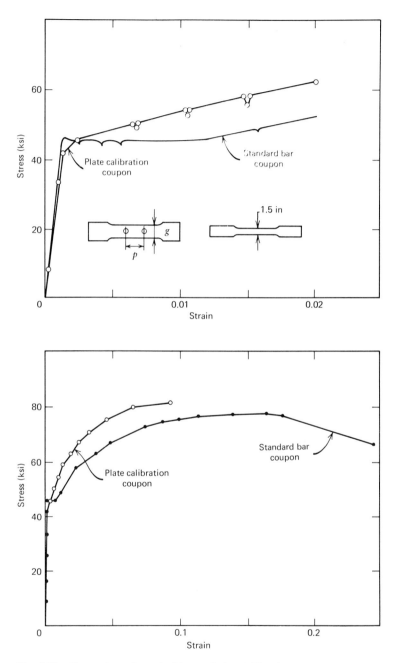

Fig. 5.24. Comparison of standard bar and plate calibration coupon.

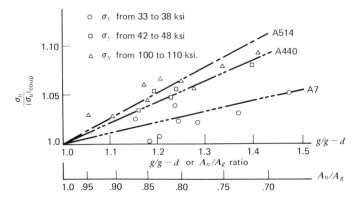

Fig. 5.25. Effect of A_n/A_g ratio on ultimate strength of tension specimen.

When a ductile metal bar is loaded and the resulting nominal stresses are plotted as a function of the strain, the characteristic relationship shown in Fig. 5.24 is observed. If a similar test is conducted on a tensile specimen with holes, the stress-strain relationship is modified, as illustrated in Fig. 5.24. For the so-called plate calibration coupon, the average strain between the two holes has been used. The ultimate strength of perforated plates at the net section is higher than the coupon ultimate strength. This results because free lateral contraction cannot develop. The increase is attributed to the "reinforcement" or bi-axial stress effect created by the holes.[5.46] As the gauge is increased, this effect is less noticeable. Figure 5.25 illustrates this behavior for different steels. The ratio $\sigma_u/\sigma_{u\,\text{coup}}$ is plotted as a function of the $g/g - d$ (or A_n/A_g) ratio. From this plot it can be concluded that a decrease in $g/g - d$ ratio (hence an increase in A_n/A_g ratio) tends to decrease the ultimate strength of the net section.

v. Effect of Type of Connected Material. The yield stress of the connected material is known to influence the ultimate strength of a joint. For a given load and resulting number of bolts, the bolt shear area is constant, whereas the net and gross areas will change depending on the type of steel used in the joint. If the connected plate carries the same load, an increase in yield stress of the plate material results in a decrease in the plate area. Since different plate areas are required the A_n/A_s ratio of a joint is affected. The influence of an increase in yield stress of the plate material on the ultimate joint strength is illustrated in Fig. 5.26. The allowable shear stress on A325 bolts is assumed to be 30 ksi, and the allowable tensile stress for the plates is taken as 22 or 30 ksi for steel with a yield stress of 36 or 50 ksi, respectively. Employing the higher strength steel reduces the

5.2 Joint Behavior After Major Slip

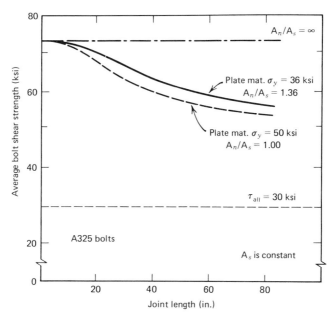

Fig. 5.26. Effect of type of connected material.

net area of the joint by a factor 22/30. Since the bolt shear area remains constant, the A_n/A_s ratio is reduced by the same factor and becomes for this particular joint equal to 1.0.

It is apparent from the comparison made in Fig. 5.26 that an increase in steel strength slightly decreases the joint strength because of the decrease in A_n/A_s ratio. The difference is not large, however, and the lower bound

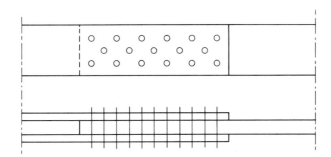

Fig. 5.27. Staggered fastener pattern.

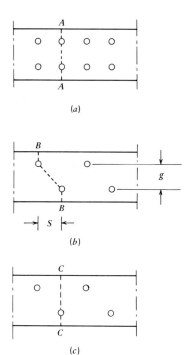

Fig. 5.28. Possible failure paths for different hole patterns.

provided by the higher strength steels can be used to develop design criteria.

vi. Fastener Pattern and Net Section Strength. Designing a tension member requires selecting a section with sufficient net area to carry the design load without exceeding the prescribed allowable stresses. Besides meeting this requirement based on strength of the connection, it is usually considered to be desirable for tension members to yield on the gross section before failure occurs at the net section. The A_n/A_g ratio reflects this requirement.

One of the parameters that influence the net area is the hole pattern. Often a simple rectangular pattern of fasteners is all that is necessary. However, in many situations a staggered hole pattern, as shown in Fig. 5.27 is required to satisfy the A_n/A_g requirement and increase the joint efficiency. For the rectangular pattern shown in Fig. 5.28a failure is likely to occur through section A-A. The reduction in area will be directly related to the diameter of the two holes. If the critical cross-section is analogous to case c, failure will occur at section C-C and the reduction in area will be

5.2 Joint Behavior After Major Slip

caused by only one hole. It is more likely that the actual failure will be bounded by these two conditions. Case *b* represents this intermediate reduction in joint capacity. The area to be deducted is a function of the stagger *s* and the gauge *g*. The following function was developed by Cochrane in 1922 and is widely used for the design of tension connections:

$$A_n = t\left(W_g - nd + \sum \frac{s^2}{4g}\right) \quad (5.11)$$

where W_g describes the gross width of the member.[5.26] With this equation the net section of a flat plate-type joint with a staggered hole pattern can be evaluated with reasonable accuracy.[5.27-5.29]

If a tension member is to yield on the gross section before failure occurs at the net section, the following equation must be satisfied

$$A_n \sigma_u \Phi \geq A_g \sigma_y \quad (5.12)$$

where σ_u and σ_y represent the tensile strength of the net section and the yield stress of the material at the gross section; Φ is a reduction factor to ensure that yielding of the gross section develops before the tensile capacity of the net section is reached. For design purposes it is convenient to express Eq. 5.12 as

$$\frac{A_n}{A_g} \geq \frac{\sigma_y}{\Phi \sigma_u} \quad (5.13)$$

It is shown in Fig. 5.25 that the tensile strength of a plate with holes depends on the A_n/A_g ratio as well as on the type of steel; for the practical range of A_n/A_g ratios the tensile strength σ_u of the net section will exceed the plate coupon tensile strength by about 7 or 8%. Consequently, using the coupon strength σ_u in Eq. 5.13 yields a conservative A_n/A_g ratio for a rectangular fastener pattern. If a staggered hole pattern is used, the net section is determined from Eq. 5.11. Since Eq. 5.11 is based on test results, the constraining effect of the hole pattern is automatically included.

To ensure that yielding on the gross section does occur before failure of the net section and also to provide a minimum factor of safety against a net section tensile failure, a reduction factor Φ is required; Φ also prevents yielding of the net section under working loads.

This examination indicates that the net section need not to be considered as the critical design section if Eq. 5.13 is satisfied. When Eq. 5.13 cannot be satisfied, the design must ensure a satisfactory margin against failure of the net section. Most of the quenched and tempered alloy steel joints do not meet the requirements of Eq. 5.13 and are to be designed on the basis of adequate net section strength.

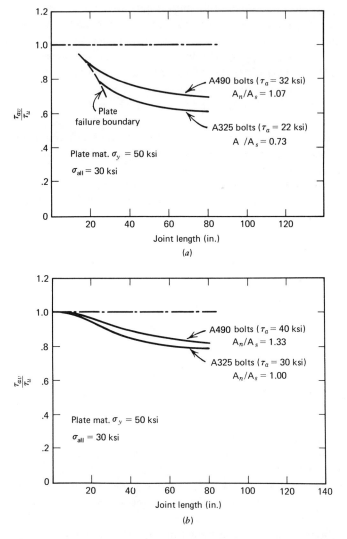

Fig. 5.29. Effect of type of fastener. (*a*) Behavior for 1973 design stresses; (*b*) behavior for proposed design stresses.

5.2 Joint Behavior After Major Slip

5.2.7 Type of Fastener

Often situations arise where the type of fastener may be variable; that is, either A325 or A490 bolts can be used. A change in bolt type corresponds to a change in A_n/A_s ratio when the net area of the joint is maintained, since the required number of bolts must change. The effect of changing the type bolt is illustrated in Fig. 5.29. Figure 5.29a corresponds to allowable bolt shear stresses of 22 and 32 ksi for A325 and A490 bolts. Figure 5.29b corresponds to allowable shear stresses of 30 and 40 ksi for the same bolts. The yield stress of the plate material was assumed to be equal to 50 ksi; this resulted in an allowable stress of 30 ksi for the plate material. By employing A490 instead of A325 bolts the bolt shear area is significantly reduced and consequently the A_n/A_s ratio is increased. The increase in A_n/A_s ratio provides a more favorable condition for the longer joints. The increase in efficiency is not as significant for shorter joints.

Besides the increase in the A_n/A_s ratio, a change from A325 to A490 bolts also reduces the joint length for a given design load. This often provides a more favorable joint condition.

5.2.8 Effect of Grip Length

Test results have shown that joints with up to 6 in. of gripped material are in close agreement with the analytical solution. Joints with larger grips and longer bolts tend to give higher ulimate loads than predicted.[5.25]

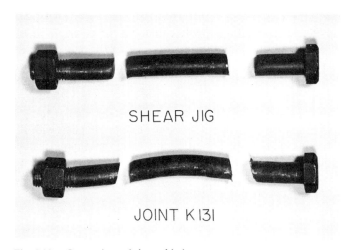

Fig. 5.30. Comparison of sheared bolts.

Fig. 5.31. Sawed sections of joints showing bolt bending.

A qualitative explanation for this observed behavior can be developed from the sheared bolts shown in Fig. 5.30. Shear tests of single bolts yield shear planes at almost 90° to the bolt axis when rigid plate elements are used, whereas the bolt from a joint with a large grip fails along an inclined shear plane. In joints fastened with long bolts the individual plates adjust to the loads they carry and the bolts assume the curved shape shown in Fig. 5.31. This results in an increased shearing area and increases the ultimate load and deformation capacity of the bolt. Hence the end fastener in a joint with long bolts deforms more than expected and permits the interior bolts to carry more load.

The extent that a bolt bends is affected by the slippage of the plates with respect to each other. Furthermore, the number of plies within the grip length of the fasteners is an important factor in developing fastener bending. For joints with high A_n/A_s ratios, the bending is more pronounced in more bolts as illustrated in Fig. 5.31. This results in an increased joint strength if failure occurs in the fasteners.

5.2.9 Bearing Stresses and End Distance

Failure of a bolted or riveted joint occurs if the applied load exceeds (*a*) the tensile capacity of the critical net section, (*b*) the shear capacity of the fas-

5.2 Joint Behavior After Major Slip

teners, or (c) the bearing strength of the material. The net section strength as well as the fastener shear strength were examined earlier. This section deals specifically with failures related to high bearing stresses on the fastener and the plate material.

After major slip has occurred in a connection one or more fasteners are in bearing against the side of the hole. A bearing stress is developed in the material adjacent to the hole and in the fastener as shown in Fig. 5.32a. Initially this stress is concentrated at the point of contact. An increase in load causes yielding and the embedment of the bolt on a larger area of contact which results in a more uniform stress distribution as indicated in Fig. 5.32b. Although the actual bearing stress distribution is not known, a uniform stress distribution can be assumed as indicated in Fig. 5.32c. The nominal bearing stress can be expressed as

$$\sigma_b = \frac{P}{dt} \tag{5.14}$$

where P denotes the load transmitted by the fastener, t the plate thickness, and d the nominal bolt diameter. Although the fastener itself is subjected to the same magnitude of compressive forces as those acting on the side of the hole tests have always shown that the fastener is not critical.[5.31, 5.32, 5.39, 5.40]

The actual failure mode in bearing depends on such geometrical factors as the end distance, the bolt diameter, and the thickness of the connected

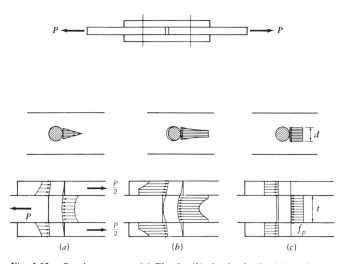

Fig. 5.32. Bearing stresses. (a) Elastic; (b) elastic-plastic; (c) nominal.

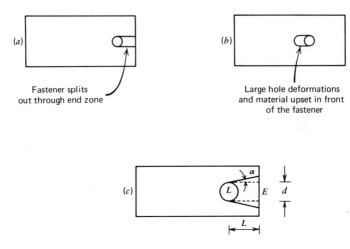

Fig. 5.33. Failure modes. (a) Fastener splits out through end zone; (b) large hole deformations and material upset in front of the fastener.

plate material. Either the fastener splits out through the end of the plate because of insufficient end distance as illustrated in Fig. 5.33a or excessive deformations are developed in the material adjacent to the fastener hole, as indicated in Fig. 5.33b. Often a combination of these failure modes will occur.

The end distance required to prevent the plate from splitting out can be estimated by equating the maximum load transmitted by the end bolt to the force that corresponds to shear failure in the plate material along the lines LE in Fig. 5.33c. The maximum shear capacity of a single bolt is equal to

$$P_s^b = mA_b\tau_u^b \tag{5.15}$$

where m is equal to the number of shear planes. The load on the fastener is also represented as

$$P_s^b = td\sigma_b \tag{5.16}$$

A lower bound to the shear resistance of part LE (Fig. 5.34c) can be expressed as

$$P_s^P = (2t)\left(L - \frac{d}{2}\right)(\tau_u^P) \tag{5.17}$$

where τ_u^P represents the shear strength (in kilopounds per square inch) of the plate material. For most commonly used steels the shear strength is

5.2 Joint Behavior After Major Slip

about 70% of the tensile strength. Hence Eq. 5.17 is transformed into

$$P_s^P = (2t)\left(L - \frac{d}{2}\right)(0.7\,\sigma_u^P) \tag{5.18}$$

where σ_u^P represents the tensile strength of the plate material, and L the edge distance of the fastener. A lower bound to the L/d ratio which will prevent the fastener from splitting out of the plate material is obtained from Eqs. 5.16 and 5.18, namely

$$\frac{L}{d} \geq 0.5 + 0.715\,\frac{\sigma_b}{\sigma_u^P} \tag{5.19}$$

This equation relates the bearing ratio σ_b/σ_u^P to the end distance represented by the L/d ratio.

Figure 5.34 compares test results with the analytical solution represented by Eq. 5.19. The experimental results were obtained from tests on symmetrical butt joints with one fastener on each side of the joint.[5.31, 5.32, 5.39, 5.40] The fastener was either a rivet or a finger-tight high-strength bolt. Figure 5.34 indicates that the test results and the analytical lower bound solution given by Eq. 5.19 tend to diverge with an increasing L/d ratio. This is expected because an increasing L/d ratio will gradually change the failure mode. For high L/d ratios, failure will not occur by shearing out the plate

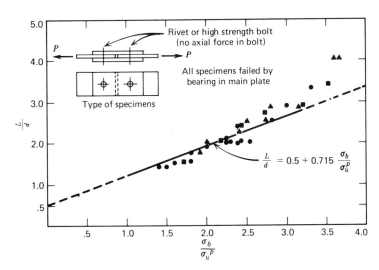

Fig. 5.34. Relationship between end distance and bearing ratio for double shear specimens. ● HSB (M16,20,24) 10K, $\sigma_u^P = 54$ ksi; □ HSB (M16,20,24), $\sigma_u^P = 74$ ksi; △ A 302 Rivets, $\sigma_u^P = 53$–76 ksi.

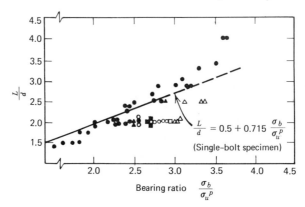

Fig. 5.35. Influence of type of specimen on the bearing ratio. ● One-bolt (or rivet) specimen (nontightened); ○ one-bolt specimen (tightened); ▲ two-bolt specimen (nontightened); △ two-bolt specimen (tightened); ■ three-bolt riveted specimen.

material in the end zone, which was assumed in the analytical solution. Failure will occur by the material piling up as indicated in Fig. 5.33b.

The test data plotted in Fig. 5.34 represent only bearing type failures of joints with one fastener on each side of the splice. Tests have shown this to be the most critical situation for the end distance.[5.39] As illustrated in Fig. 5.35, joints with one fastener require a larger L/d ratio than joints having more than one fastener in a line. It is also apparent that providing a clamping force in the bolt leads to an increase in the ultimate bearing ratio. This indicates that the load is partially transmitted by frictional resistance on the faying surfaces. Consequently the real bearing stress is less than the "ultimate" bearing stress computed on the basis of the total applied load.

All the data summarized in Figs. 5.34 and 5.35 were obtained from tests on symmetric butt joints. Failure always occurred in the main plate. If the lap plates are relatively thin compared to the main plate, then failure may occur in the lap plates. Test results have indicated that in these situations bearing failures are influenced by "catenary action," which causes bending in the lap plates.[5.31, 5.40] The thin lap plates bend outwards and decrease the ultimate bearing strength of the connection.

5.3 JOINT BEHAVIOR UNDER REPEATED LOADING

5.3.1 Basic Failure Modes

The behavior of a bolted connection under repeated loading is directly influenced by the type of load transfer in the connection. The applied load

5.3 Joint Behavior Under Repeated Loading

can be transferred either by friction on contact surfaces, by shear and bearing of the bolts, or by both depending on the direction of the applied load, the magnitude of the clamping force, the condition of the faying surfaces, and the possible occurrence of major slip. Tests have shown that each load transfer mechanism develops its own characteristic failure pattern under repeated loadings.[5.18] These characteristic conditions are best explained and illustrated by examining the stress distribution throughout the joint.

Figure 5.36 shows schematically an idealized lap joint subjected to a cyclic, in plane force. Assuming that no major slip occurs, hence that the external load is completely transmitted by friction on the faying surfaces, implies a high concentration of shear stresses at point A in Fig. 5.36. This results from the large differences in strain between the lap and main plates. The interface would be required to transmit a highly concentrated shear force at A, if it were not relieved by microslip at that point. In many tests it was observed that under these conditions crack initiation and growth usually occurred in the gross section, in front of the first bolt hole as indicated schematically in Fig. 5.36a. The cracks initiated on the faying surfaces of the connected plates. This phenomena is often referred to as fretting and occurs at the interface between metallic surfaces that are in contact and

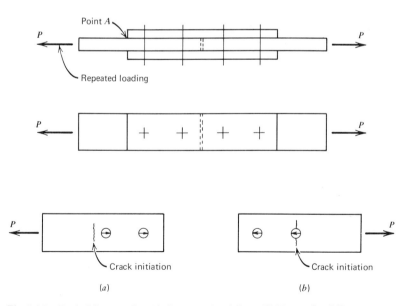

Fig. 3.36. Basic failure modes. (a) Gross section failure; (b) Net section failure.

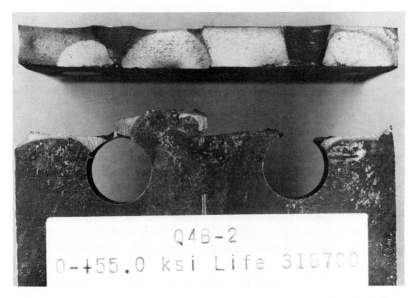

Fig. 5.37. Typical fretting-type failure in gross section. (Courtesy of University of Illinois.)

which slip minute amounts relative to each other under the action of an oscillating force.[5.41] Even the small relative displacements between the lap and the main plates at point A (see Fig. 5.36) may be sufficient to initiate a fretting failure. The obvious effect of fretting is to damage the faying surfaces. Stress concentrations are also introduced which in many cases lead to crack initiation and a further reduction in fatigue strength.

Tests have indicated that high contact pressures only exist in a small area around the bolt hole.[5.47, 5.48] The normal stress due to the clamping force decreases rapidly from a maximum condition at the edge of the hole. The region where the normal stress acts depends on such geometrical factors as the plate thickness and bolt diameter. Usually the circular pressure area falls within twice the diameter of the bolt. For this reason the crack initiates at a section between the end of the lap plate and the bolt hole where the combination of microslip and normal pressure is more critical.

A typical fretting failure is shown in Fig. 5.37. Discontinuities of the mill scale, the effective clamping zone of the bolt and the frictional resistance all influence the point where fretting is initiated. Fretting is often apparent during fatigue testing. A powdery rust and mill scale dust usually works out from between the plates during testing.[5.20]

Fig. 3.38. Failure at net section of bearing-type joint. (Courtesy of University of Illinois.)

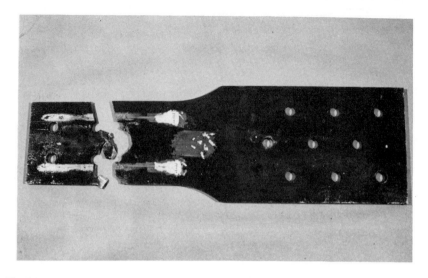

Fig. 5.39. Crack initiation and growth at net section due to fretting. (Courtesy of U.S. Steel Corp.)

The other major type of fatigue failure that occurs in bolted or riveted shear type splices is illustrated in Fig. 5.36b. The crack initiates at the edge of the hole and grows in the region of the net section. This condition occurs when most of the load is transmitted by shear and bearing, a situation which frequently develops in joints where the applied load exceeds the slip resistance of the faying surfaces. This results in higher net section stresses and the edge of the hole becomes the point of crack initiation. Failure is brought about by fracture of the net section, as shown in Fig. 5.38.

Both types of failure have been observed in tests. Often the two types of failure occur simultaneously in the same joint as illustrated in Fig. 5.39. Final failure occurs partly through the net section and partly through the gross section.

Besides the bolt relaxation normally experienced after installation, some additional relaxation (5%) was observed during cyclic loading.[5.18] Tests have indicated that the total loss of bolt tension was rarely more than 10% of the initial bolt tension.[4.9, 5.18, 5.20]

5.3.2 Fatigue Strength of Bolted Butt Joints

The stress-life relationship is best described by a logarithmic transformation of cycle life and maximum stress or stress range.[2.5-2.7] Therefore, data from fatigue tests are generally described using the relationship

$$\log N = A + B \log S \qquad (5.20)$$

5.3 Joint Behavior Under Repeated Loading

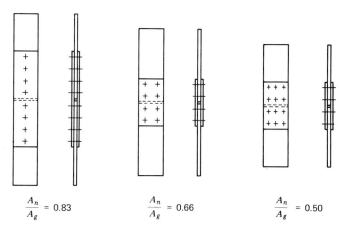

Fig. 5.40. Test specimen.

where N represents the number of cycles, S the maximum stress or stress range, and A and B are constants. Plotted on a log-log scale, Eq. 5.20 results in a straight line. Recent work on welded details has suggested that stress range is the major stress variable.[5.51]

Since two basic types of crack growth were observed in bolted joints, one in the gross section and the other in the net section, the test results have been correlated with the stresses associated with both areas. If no major slip developed during the life of a specimen and high clamping forces are present, failure occurs in the gross section. Therefore, an examination of

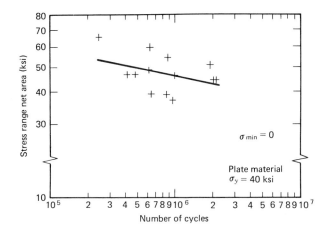

Fig. 5.41. Experimental S-N curve based on net area stress.

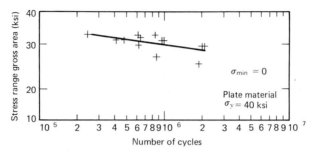

Fig. 5.42. Experimental S-N curve based on gross area stress.

the test data using the stress range on the gross area seems reasonable. Net section stresses depend on geometrical factors such as the arrangement of the bolts in the joint. This causes large variation in stress and is partly responsible for the large scatter in test data when gross section failures are correlated on the basis of net section stresses. This is illustrated in Fig. 5.41 where test results from three different types of joints are compared.[5.5] The joint geometry is given in Fig. 5.40. Major slip did not occur because of the design conditions. Nearly all failures were through the gross section area. The test data indicate substantial variation in fatigue strength for the three different geometrical conditions. Figure 5.42 shows the same data plotted on the basis of the gross section stresses. These figures illustrate that the use of the gross area decreased the scatter in the test results significantly.

Major fatigue work on bolted and riveted connections was performed at the University of Illinois,[4.9, 5.20, 5.42] Northwestern University,[3.6, 3.7, 5.19] and in Germany.[5.18] Figure 5.43 shows some results of tests on bolted slip resistant joints subjected to repeated loading. Since major slip did not occur in these joints, failure was caused by crack growth in the gross section. Therefore, the gross section area was used to determine the stress range, S, when evaluating the available test data. Most of the data were obtained from tests on steel specimens with a yield stress between 34 and 60 ksi. Data are available on joints fabricated from quenched and tempered alloy steel (A514) as well.[5.20] The yield stress (taken as the 0.2% offset) of the A514 steel was about 120 ksi. Although the data plotted in Fig. 5.43 show considerable scatter, it indicates that the yield stress of the material does not significantly influenc the fatigue behavior of bolted joints.

Fatigue tests on slip resistant joints in which the applied load on the specimen was reversed ($R < 0$), are shown in Fig. 5.44. The stress range includes the full compressive portion of the stress cycle. A comparison between the data plotted in Figs. 5.43 and 5.44 indicates that for a given stress range, a slightly higher life was observed for the specimens subjected

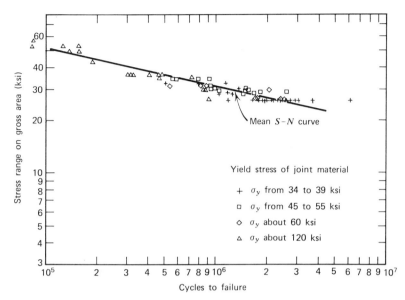

Fig. 5.43. Test results for slip-resistant joints ($R = 0$).

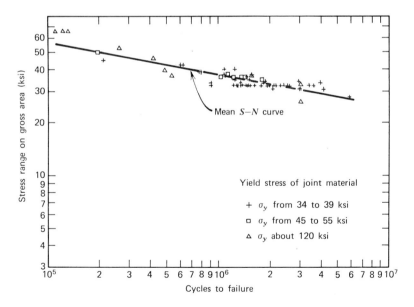

Fig. 5.44. Test results for slip-resistant joints ($R < 0$).

to stress reversal condition as compared to the zero to tension ($R = 0$) specimens. This seems reasonable, since recent crack growth studies have indicated that when residual tensile stresses are not present, the compression stress cycle is not as effective in extending the crack as the tensile component.[5.43] Considering the full stress range effective is a conservative estimate of the fatigue strength of bolted joints.

Joints with low slip resistance as a result of less clamping force or low slip coefficients are subjected to higher stresses on the net section, when the slip resistance of the joint is exceeded by the applied load. When subjected to repeated loading conditions, crack initiation and growth occurs in the net section of such joints. Consequently, their performance under these loading conditions is related to the magnitude of stresses on the net section.

If the slip resistance of a joint is exceeded, the connection slips into bearing and the applied load is transmitted partly by shear and bearing on the fastener as well as friction on the faying surfaces. Tests have indicated that the fatigue life determined from a plate with a hole provides a lower bound estimate of the fatigue strength of bolted joints which have slipped into bearing.[5.18, 5.19] The improved behavior of bolted joints as compared to the plate specimens with a hole is primarily attributed to the influence of the clamping force in the fasteners. A more favorable stress condition exists in the joint because part of the load is transmitted by friction on the faying surfaces.

All available test results on bolted joints fabricated from steels with yield stress varying from 36 to 120 ksi are plotted in Fig. 5.45. The stress range used to plot the test data was computed on the basis of the net or gross section area, depending on whether or not joint slip occurred. It is apparent in Fig. 5.45 that both bearing type and slip-resistant joints subjected to reversal-type loading provide higher fatigue strength.

The data plotted in Fig. 5.45 show a significant scatter even within the individual categories of joint types and loading conditions. This is mainly attributed to the fact that the data originated from various sources and reflected the variability in the hole fabrication, bolt clamping force, joint configuration, and other variables. Also, different tightening techniques were used to install the fasteners and this may have resulted in significant variations in clamping forces of the fasteners. These variations as well as differences in joint geometry and hole preparation used in the various test series tend to increase the natural scatter of the data.

Only a few test data are available on specimens subjected to a tension-to-tension stress cycle.[5.20, 5.42] Except for a few tests on A514 steel joints, most of the test data were acquired at stresses that exceeded the yield point on the net section and often approached or exceeded the yield point on the

5.3 Joint Behavior Under Repeated Loading

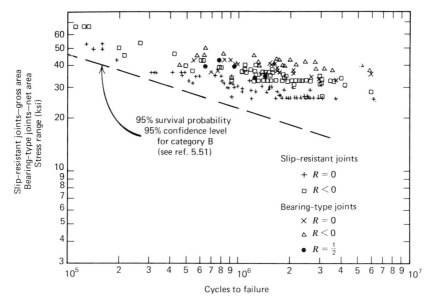

Fig. 5.45. Summary of test results of bolted joints.

gross section. The data from A514 steel joints are in good agreement with the results in Fig. 5.45. The data from specimens that exceeded the yield point by such large margins were not considered as they are not representative of the conditions that occur in actual structures.

Figure 5.45 also indicates that most of the data are concentrated in a stress range band between 25 and 40 ksi. Additional information in the short life region and for very large number of cycles are needed for a better understanding of the fatigue strength in these ranges. Despite these shortcomings, the data presented in Fig. 5.45 can be used to develop conservative recommendations for the design of bolted joints subjected to repeated type loading.

For design purposes the data on bolted joints was compared with the 95% confidence limits used to define category B for proposed fatigue specifications.[5.51] Although category B was derived from tests on plain welded beams, it was apparent that the proposed design relationship provided a reasonable lower bound to the test data on bolted joints. The use of this lower bound for bolted joints results in conservative design relationships. It is apparent that slip-resistant joints designed on the basis of gross section and bearing-type joints designed on their net section provide about the same fatigue strength.

The test data shown in Figs. 5.43 to 5.45 was developed from symmetric butt joints with a maximum of three fasteners in a line parallel to the direction of the applied load. Only a few test results of longer joints have been reported.[5.44] Tests of high-strength bolted joints with two, four or six fasteners in a line indicated no significant influence of the number of bolts on the fatigue strength. These tests did show that the frictional resistance of the faying surfaces does affect the fatigue strength. An increase in slip resistance improved the fatigue behavior.

5.4 DESIGN RECOMMENDATIONS

5.4.1 Introduction

The mathematical model presented in Section 5.2 provides a reasonable prediction of joint behavior at working loads as well as at the ultimate strength level. However, it is not suitable for design. Current design practice treats mechanically fastened joints on the basis of allowable stresses acting on either the gross or net area of the member and the average shear stresses on the fasteners. This design shear stress concept for fasteners has been used for over a century. All fasteners are assumed to carry an equal share of the load. Once the member forces are known from a structural analysis, the required number of fasteners can be determined directly on the basis of the allowable shear stress permitted on the fastener. Hence the load transmitted by a bolted joint with n fasteners and m possible shear planes per bolt through the bolt shank can be expressed as

$$P = mn\tau_b A_b \qquad (5.21)$$

where τ_b represents the shear stress on the fastener and A_b the nominal bolt area. If the shear planes pass through the threaded part of the bolt, Eq. 5.21 is transformed into

$$P = 0.75 mn\tau_b A_b \qquad (5.22)$$

as discussed in Section 4.10.

Although the bolts in a slip-resistant joint are not actually subjected to shearing forces, it is convenient to account for the joint load in terms of an equivalent shear stress on the nominal bolt area. The load transmitted by a slip-resistant joint can be expressed in the form of Eq. 5.21. In this case the shear stress τ_b is a fictitious stress, since the load is actually transferred by the frictional resistance on the faying surfaces.

Design criteria for connections can be based upon performance and strength. In a slip-resistant joint unsatisfactory behavior would result if major slip occurred. The function of the structure may be impaired due to misalignment or other unsatisfactory conditions that may result. In most

5.4 Design Recommendations

bolted joints, the shear stress on the fastener or the tensile stress on the net or gross area is the factor governing design. Minor slip is not critical to the joint's performance and strength is the major factor that governs the design.

The ultimate capacity of both slip-resistant and other bolted joints is limited by failure of one or more components of the joint. Joint strength provides an upper bound for all joint types. Hence allowable design stresses for slip-resistant joints can at best equal the allowable stresses permitted in other bolted joints. In other words, to design a slip-resistant joint, the slip resistance of the joint is determined on the basis of factors such as the surface condition, the bolt type, the tightening procedure, the number of bolts, and the number of slip planes. The slip resistance is then compared with the bolt shear capacity of the joint based upon the number of shear planes per bolt and their location (through the shank or through the threaded part of the bolt) and the number of bolts in the joint as well as the bolt quality. The smaller value of either the shear strength and slip resistance is governing.

In load factor design the members are selected so that the structure reaches its maximum strength at the factored load. The factored load is determined by multiplying the working load by a factor which is greater than 1.0 (see Section 2.4).[5.50] Obviously, it is necessary for the joint to have a reserve margin of strength at the working load level. To ensure serviceability, consideration must be given to the control of deflections, deformations and fatigue of the structure at its service or working load level. This requires the connection to be designed for strength and then checked for performance under working load conditions.

5.4.2 Design Recommendations—Fasteners

i. Allowable Stress Design Bolted Joints. The balanced design concept has been used to develop design criteria for mechanically fastened joints in the past. This design philosophy results in wide variations in the factor of safety for the bolt, because the ratio of the yield point to the tensile strength changes with various types of steel.[5.49] Current (1972) specifications[1.4] provide ratios of tensile strength to allowable tensile stress equal to 2.64, 2.48, and 2.00 for A36, A440, and A514 steel. Furthermore, the balanced design concept has no meaning when applied to long joints, because the end fasteners may "unbutton" before the plate material can attain its full strength or before the interior bolts can be loaded up to their full strength. This "long joint" effect depends on the type of joint material as well as on the type of fastener.

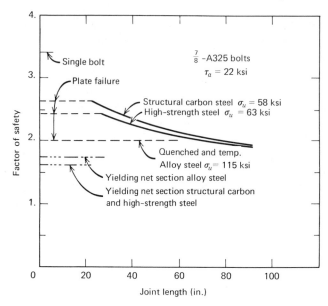

Fig. 5.46. Factor of safety versus joint length for A325 bolts.

All of these factors result in a variable factor of safety as illustrated in Fig. 5.46. The factor of safety against failure of the joint is plotted as a function of joint length for several steels fastened with A325 bolts. An allowable shear stress of 22 ksi was used to proportion the fasteners. The allowable tensile stress on the net section of the joint was taken as 60% of the yield stress or 50% of the tensile strength of the plate material, whichever was smaller. It is apparent that a different approach is desirable; one that will provide both a rational method of determining the allowable stresses and a uniform, or at least a more consistent factor of safety. It appears that a more logical criterion to establish allowable stresses for the fasteners should consider the fastener strength over the full range of joint behavior.

To determine the magnitude of the factor of safety deemed adequate for the fasteners, two aspects can be considered: (1) what the factor of safety has been in the past, and (2) what it ought to be. If past practice is studied for riveted or bolted structural carbon steel joints, the factor of safety against shear failure of the fastener is found to vary from approximately 3.3 for compact joints* to approximately 2.0 for joints with a length in

---

* A compact joint is defined as a joint in which the average fastener shear stress at the ultimate load level is equal to, or almost equal to, the shear strength of a single fastener. The "unbuttoning" effect is negligible in these joints.

5.4 Design Recommendations

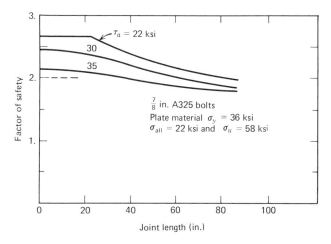

Fig. 5.47. Factor of safety for structural carbon steel joints fastened by A325 bolts.

excess of 50 in. This is illustrated in Fig. 5.46 for A325 bolts. The lower factor of safety for the longer joints has apparently been adequate in the past. In fact, according to past practice, the largest and often most important joints have probably had the lowest factor of safety. Experience has shown that this factor of safety has provided a safe design condition. This indicates that a minimum factor of safety of 2.0 has been satisfactory; the same margin is also used for fasteners in tension. Minimum specified

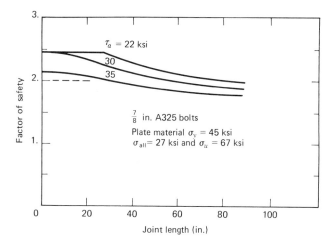

Fig. 5.48. Factor of safety for high-strength steel joints fastened by A325 bolts.

mechanical properties of the bolt and plate material were used to determine these lower bound conditions. Materials actually used as components of the joint may well provide strengths that exceed minimum specified properties. This results in an increased factor of safety. Note that a minimum factor of safety equal to 2.0 for bolts in shear is not only in line with the factor of safety presently (1973) used for bolts in tension, but the same factor of safety against ultimate is also provided by quenched and tempered alloy steel tension members.[2.11]

In Fig. 5.47 the factor of safety is plotted as a function of the joint length for different allowable shear stresses in $\frac{7}{8}$-in. A325 bolts, installed in structural carbon steel with a yield stress of 36 ksi and a tensile strength of 58 ksi. Joint length is defined as the length required to transfer the load from the main plate into the splice plates. Hence for a symmetric butt splice, the joint length is equal to half the total length of the lap plate. For a single lap joint it is equal to the overall length of the joint. Figures 5.48, 5.49, and 5.50 show plots for other combinations of plate material and bolt grades. A minimum factor of safety of 2.0 is provided when a 30-ksi allowable shear stress is used for A325 bolts installed in structural carbon steel up to a joint length of 60 in. High-strength steel with a tensile strength of 66 ksi and fastened by A325 bolts provides a minimum factor of safety of 2.0 up to about 50 in. Figures 5.49 and 5.50 show that a 40-ksi allowable shear stress for A490 bolts would provide the needed margin for lengths up to about 50 in. For joints with a length exceeding 50 in. the allowable shear stress must be reduced to ensure a minimum factor of safety of 2.0. A 20% reduction in

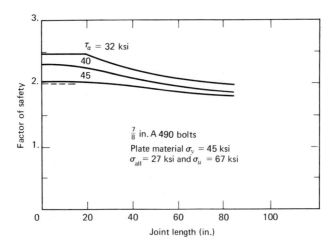

Fig. 5.49. Factor of safety for high-strength steel joints fastened by A490 bolts.

5.4 Design Recommendations

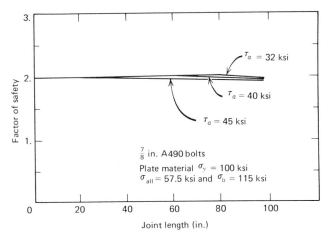

Fig. 5.50. Factor of safety for quenched and tempered alloy steel joints fastened by A490 bolts.

the allowable shear stress provides this margin for joint lengths between 50 and 90 in. as illustrated in Figs. 5.47 to 5.50.

DESIGN RECOMMENDATIONS BOLTED JOINTS

Allowable Stress Design

Shear Stresses for High-Strength Bolts

$$\tau_a = \beta_1 \beta_2 \beta_3 \tau_{\text{basic}}$$

where
$\tau_{\text{basic}} = 30$ ksi — A325 bolts
$\tau_{\text{basic}} = 40$ ksi — A490 bolts

When joints are not slip-resistant:

$$\beta_1 = \beta_2 = \beta_3 = 1.0$$

if joint length exceeds 50 in.:

$$\beta_3 = 0.8$$

Allowable Joint Loads

(a) Shear planes pass through bolt shank

$$P = mn\tau_a A_b$$

(b) Shear planes pass through bolt threads

$$P = 0.75 mn\tau_a A_b$$

ii. Load Factor Design Bolted Joints. In load factor design the connections and structural members are proportioned so that the product of maximum strength and a reduction factor Φ is at least equal to the applied design loads multiplied by their respective load factors. The reduction factor Φ is introduced to assure that the maximum strength of a structure is limited by the capacity of its members rather than by premature failure of the connections. The Φ factor also accounts for the variability in strength of a connection. A uniform Φ factor of 0.75 has been suggested for mechanical fasteners loaded in shear.[5,50]

The shear strength of a single fastener is about 60% of its tensile strength (see Section 4.2). A Φ factor of 0.75 yields shear stresses comparable to those obtained by factoring the suggested working allowable shear values by 1.8. The same Φ factor is applicable to A307 bolts and to A502 rivets. The ultimate shear capacity of a high-strength bolted connection is affected by the location of the shear planes. If a plane intersects the bolt threads, only the root area is effective in resisting the shear. This reduces the joint shear capacity by about 25% (see Section 4.10).

DESIGN RECOMMENDATIONS BOLTED JOINTS

Load Factor Design—Shear Loading

$$\text{Design strength} = \Phi F$$

where F — average shear strength = $0.60\sigma_u$
Φ — reduction factor = 0.75
If joint length exceeds 50 in. $\Phi = 0.60$

Factored Joint Loads
(a) Shear planes pass through boltshank

$$P = mn\Phi F A_b$$

(b) Shear planes pass through bolt threads

$$P = 0.75 mn\Phi F A_b$$

iii. Slip-Resistant Joints. By assuming equal clamping forces throughout a joint, the slip resistance of a connection is given by

$$P_s = mn T_i k_s \qquad (5.23)$$

For a given joint geometry the slip resistance is directly proportional to the product of the initial clamping force T_i and the slip coefficient k_s. Both

5.4 Design Recommendations

quantities have considerable variance that must be considered when determining design criteria for slip-resistant joints. Since both the frequency distributions for k_s and T_i are known for different surface conditions, bolt types and tightening procedures (see Sections 5.1.5 and 5.1.6), the joint frequency distribution for the product $k_s T_i$ can be determined.[5.33] The product $k_s T_i$ can be transformed into a shear stress on the nominal bolt area by equating the slip resistance to an equivalent shear.

Hence

$$mn\tau_b A_b = mn k_s T_i \qquad (5.24)$$

where τ_b is an equivalent shear stress and A_b the nominal bolt area. If

$$\alpha = \frac{T_i}{T_{i\ \text{spec}}} \qquad (5.25)$$

and

$$T_{i\ \text{spec}} = 0.7 \, A_s \sigma_{u\ \text{spec}} \qquad (5.26)$$

where A_s represents the stress area of the bolt, then Eq. 5.24 can be expressed as

$$\tau_b = 0.7 \, k_s \alpha \frac{A_s}{A_b} \sigma_{u\ \text{spec}} \qquad (5.27)$$

The ratio of the stress area to the nominal bolt area varies from 0.736 for a ⅝-in.-diameter bolt up to 0.774 for a 1-in. bolt. An average value of 0.76 was selected. The minimum specified tensile strength for A325 bolts in sizes ½ through 1 in. is 120 ksi. Substituting these values into Eq. 5.27 yields

$$\tau_b = 63.8 k_s \alpha \qquad (5.28)$$

Equation 5.28 relates the equivalent shear stress on the fastener to the known parameters α and k_s (as described in Section 5.1). The frequency distribution curve of the equivalent shear stress τ_b can be evaluated from Eq. 5.28 and is shown in Fig. 5.51a. This particular curve is based on an α coefficient corresponding to A325 bolts installed by the turn-of-nut method. The surface condition of the joint, characterized by the slip coefficient k_s, is assumed as a clean mill scale condition. The cumulative frequency (see Fig. 5.51b) gives the nominal shear stress on the bolt as a function of the slip probability. For a given slip probability, the equivalent shear stress can be evaluated. Similar curves can be constructed for other surface conditions and bolt grades. The results of such an analysis are summarized in Table 5.2 for different surface conditions, bolt grades and slip probabilities. The maximum allowable shear stresses assuming 95%

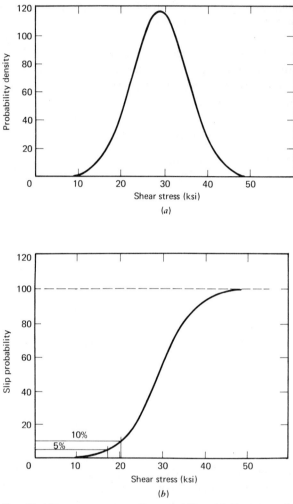

Fig. 5.51. Allowable shear stress versus slip probability. (*a*) Frequency distributions; (*b*) cumulative frequency curve.

confidence limit and 90% confidence are given as a fraction (equal to β_1) of basic allowable shear stress for the bolt (i.e., 30 and 40 ksi for A325 and A490 bolts, respectively). The number in brackets represents the allowable shear stress in ksi when $\beta_2 = \beta_3 = 1.0$.

In evaluating the allowable shear stresses for A325 bolts, the specified minimum tensile strength was presumed to be 120 ksi. The specified tensile strength for A325 bolts in sizes over 1 in. is 105 ksi. Experience has shown

5.4 Design Recommendations

that the actual strength of A325 bolts over 1 in. diameter usually ranges from 20 to 34% above the minimum specified tensile strength. Furthermore, the A_s/A_b ratio for these sizes is about 0.81 as compared to the value 0.76 for sizes less than 1 in. diameter. An increase in the A_s/A_b ratio increases the shear stress as is apparent from Eq. 5.27. Hence the shear stresses listed in Table 5.2 are assumed applicable to all commonly used bolt sizes.

The allowable shear stresses listed in Table 5.2 are applicable to bolts installed by the turn-of-nut method. In Section 5.1.6 it was shown that the calibrated wrench method provided a lower clamping force than the turn-of-nut method. This reduces the slip resistance of a joint. To account for this a reduction factor β_2 is introduced. On the basis of available information, as illustrated in Fig. 5.8 from Chapter 5.1, a value of $\beta_2 = 0.85$ appears reasonable for both A325 and A490 bolts.

A reduction factor β_3 was introduced to count for the effect of fabrication factors on the slip resistance of joints; for example, depending on the amount of oversize of the hole or the direction of the slotted holes with respect to the expected slip direction, a reduction in slip resistance may result. Chapter 9 deals specifically with oversize and slotted holes and discusses in greater detail the influence of these fabrication factors on the slip resistance of a joint.

Table 5.2. Reduction Factors β_1 for Evaluation of Design Shear Stresses for Slip-Resistant Joints[a]

	Probability of Slip			
	A325 Bolts		A490 Bolts	
Surface Treatment	5%	10%	5%	10%
A36, A440 clean mill scale	0.59	0.68	0.51	0.59
	(17.8)	(20.3)	(20.7)	(23.6)
A36, A440 grit blasted	0.99	1.09	0.87	0.97
	(29.8)	(32.7)	(34.8)	(38.2)
A36, A440 semi-polished	0.55	0.61	0.48	0.53
	(16.7)	(18.3)	(19.4)	(21.3)
A514 grit blasted	0.69	0.75	0.60	0.65
	(20.7)	(22.5)	(24.3)	(26.2)

[a] Summary β_1 factors: The number in parenthesis represents allowable shear stress for $\beta_2 = 1.0$ (turn-of-nut method) and $\beta_3 = 1.0$. Coated surfaces are given in Chapter 12.

Performance as well as strength must be considered in the design of slip-resistant joints. As briefly mentioned in Section 5.4.1, to provide a uniform minimum factor of safety with respect to the ultimate load, the allowable load of a slip-resistant joint should not exceed the allowable load of an identical joint based on strength criteria only. In other words, the maximum allowable load for a joint, based on strength considerations, forms an upper bound for the allowable load that must not be exceeded by a slip-resistant joint. This condition only affects the design of slip-resistant joints with high slip resistance. For example, the design load based upon strength may be reduced by slip planes through the threaded portion of the bolts. If this is less than the load permitted on a slip-resistant joint, the strength criteria becomes the governing factor. This requires that the allowable shear stress for a slip-resistant joint does not exceed the basic shear stress of 30 or 40 ksi for A325 or A490 bolts, respectively. It can be seen in Table 5.2 that such a situation arises in a joint with grit-blasted surfaces, fastened by A325 bolts, and a 10% probability of slip. Assuming the reduction factors β_2 and β_3 are equal to 1.0 yields a design stress of 32.7 ksi. Strength criteria would limit the maximum shear stress to 30 ksi for A325 bolts. If the joint were fabricated with oversize or slotted holes, the reduction factor β_3 would decrease to about 0.70. Consequently, the allowable design shear stress becomes 70% of 32.8 ksi that governs.

DESIGN RECOMMENDATIONS FOR SLIP-RESISTANT JOINTS

$$\tau_a = \beta_1 \beta_2 \beta_3 \tau_{basic}$$

where τ_{basic} 30 ksi for A325 bolts
τ_{basic} 40 ksi for A490 bolts
β_1 depends on slip probability
        surface treatment } See Table 5.2
        type of bolt
β_2 depends on tightening procedure
        turn-of-nut method $\beta_2 = 1.0$
        calibrated wrench method $\beta_2 = 0.85$
β_3 depends on fabrication factors.
        e.g., oversize or slotted holes $\beta_3 = 0.70$

To ensure minimum factor of safety with respect to ultimate load for slip-resistant joints,

$$mn\tau_a A_b \leq \bar{P}$$

where $\bar{P}$ is the design load of connection based on strength criteria only (see Section 5.4.2i)

5.4 Design Recommendations

5.4.3 Design Recommendations—Connected Material

i. Static Loading. It was shown in Section 5.2.6 that for tension members to yield on the gross section the ratio of the net to gross section area must satisfy

$$\frac{A_n}{A_g} \geq \frac{\sigma_y}{\Phi \sigma_u} \qquad (5.29)$$

The factor Φ is smaller than 1.0 and ensures that yielding will occur on the gross section area before the tensile capacity of the net section is reached. The minimum specified mechanical properties provide a σ_y/σ_u ratio for carbon steel equal to approximately 0.62. If Φ were unity the minimum A_n/A_g ratio for carbon steel is 0.62. With the allowable stress on the gross section of a carbon steel tension member equal to 22.0 ksi and the A_n/A_g ratio equal to 0.62, a net section stress equal to 35.4 ksi would result. This would also provide a factor of safety with respect to the ultimate load of 1.64. To provide a minimum factor of safety of 2.0 (as provided by the fasteners), a Φ factor of 0.85 is required for all types of structural steel to permit the design to be based on the gross area alone. Hence a tension member may be designed on the basis of the gross section if

$$\frac{A_n}{A_g} \geq \frac{\sigma_y}{0.85 \, \sigma_u} \qquad (5.30)$$

The minimum A_n/A_g ratios as given by Eq. 5.30 yields net section stresses equal to 30 ksi for structural carbon steel members with a yield stress of 36 ksi. For high-strength steel members with a yield stress of 42 ksi, the maximum net section stress is equal to 34 ksi. These stresses have a very local character and do not influence the behavior of the connected member. An analogous provision has been used for bending members for some time. Design can be based on the gross area provided no more than 15% of flange area is removed.

When the net area to gross area ratio is less than specified by Eq. 5.30, the stress on the net section of the tension member must not exceed 50% of the tensile strength of the material.

DESIGN RECOMMENDATIONS NET SECTION UNDER STATIC LOADING CONDITIONS

When

$$\frac{A_n}{A_g} \geq \frac{\sigma_y}{0.85 \, \sigma_u}$$

the gross section of a tension member can be used to proportion the member. When

$$\frac{A_n}{A_g} < \frac{\sigma_y}{0.85\,\sigma_u}$$

the tensile stress on the net section should not exceed $\frac{1}{2}\sigma_u$ hence

$$\sigma_a \leq 0.5\,\sigma_u$$

ii. Repeated Loading. Results of fatigue tests on slip-resistant as well as other types of bolted joints were discussed in Section 5.3. It was shown that the type of failure was related to the manner in which the applied load was carried by the joint. If transmitted by frictional resistance on the contact surfaces alone, failure was through the gross section. When slip occurred and part of the load was transmitted by bearing and shear, failure generally occurred through the net section. The fatigue strength of the gross section of slip-resistant joints was about equal to the fatigue strength of the net section of joints that had slipped into bearing under nonreversible loading.

Design category *B* which was derived from tests on plain welded beams[5,51] provides a reasonable lower bound estimate for the stress range-life relationship of bolted joints. The allowable stress ranges, determined from this stress range-life relationship for different loading conditions, are summarized in Table 5.3. A stress range of 16 ksi was estimated for a life of 2 million cycles or more.

For the design of high-strength bolted joints under cyclic loading the suggested stress range can be applied to: (*a*) the gross section area of slip-resistant joints with a slip probability of 5% or less, and (*b*) the net section area for other bolted joints. This provides design stresses for clean mill scale conditions that are in reasonable agreement with current practice. Joints subjected to reversal of stress should always be designed as slip resistant joints to prevent excessive movement of the connected parts.

The stress range on the net section area governs the design of bolted joints that have a slip probability greater than 5%. These joints should not be used in situations where reversal of load occurs. However, slip in the direction of the maximum applied load is not critical unless the load is reversed.

Application of the stress ranges given in Table 5.3 provides a conservative design for both slip-resistant and bearing-type bolted joints. Better estimates of the stress range-life relationship may be developed when additional experimental data becomes available.

5.4 Design Recommendations

Table 5.3. Allowable Range of Stress for the Plate Material

Design Load Cycles		Stress Range for 95% Survival (ksi)
From	To	
20,000–100,000		45.0
100,000–500,000		27.5
500,000–2,000,000		18.0
Over 2,000,000		16.0

DESIGN RECOMMENDATIONS FOR JOINT MATERIAL UNDER REPEATED LOADING

Slip-Resistant Joints

Stress range on gross section area governs if the slip probability is less than or equal to 5%.

Other Bolted Joints

Stress range on the net area governs if the slip probability is greater than 5%. Stress reversal is not permitted. Allowable stress range for both types is given in Table 5.3.

iii. **Bearing Stresses.** Section 5.2.9 showed that the lower bound L/d ratio which prevents a single fastener from splitting out of the plate material can be expressed as:

$$\frac{L}{d} \geq 0.5 + 0.715 \frac{\sigma_b}{\sigma_u{}^P} \tag{5.31}$$

Butt joints with a single fastener were more critical than joints with multiple fasteners in a line. The clamping force in a high-strength bolt also has a favorable influence on the bearing strength of the connection. Hence design recommendations based on test results of finger-tight single fastener specimens provide a conservative estimate of the required end distance.

The test results indicate that Eq. 5.31 provides an acceptable lower bound solution to the strength of the end zone for L/d ratio up to 3.0 as illustrated in Fig. 5.52. When the L/d ratio exceeds 3.0 the failure mode changes gradually from a "shearing-type" failure to one in which large hole and material deformation occurs.

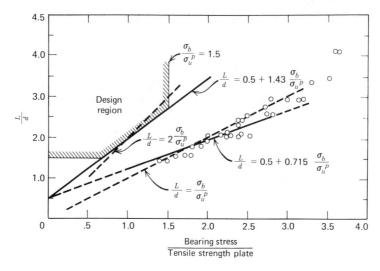

Fig. 5.52. Comparison design recommendations for allowable stress design and test results.

An alternate relationship can be developed considering a simpler expression directly relating the L/d ratio to the bearing stress-tensile stress ratio. This yields

$$\frac{L}{d} \geq \frac{\sigma_b}{\sigma_u^P} \qquad (5.32)$$

This relationship is also plotted in Fig. 5.52 and is in good agreement with the test data.

a. Allowable Stress Design. If a minimum factor of safety with respect to ultimate load of 2.0 is selected, the required L/d ratio becomes

$$\frac{L}{d} \geq 0.5 + 1.43 \frac{\sigma_b}{\sigma_u^P} \qquad (5.33)$$

As is shown in Fig. 5.52, Eq. 5.31 defines the L/d ratio up to a bearing stress-tensile strength ratio of 3.0. The suggested factor of safety of 2.0 against bearing failure is comparable to the factors of safety against shear or tension failure of the fasteners and the tensile strength of the net section.

If the alternate formulation is used the required L/d ratio becomes:

$$\frac{L}{d} \geq 2 \frac{\sigma_b}{\sigma_u^P} \qquad (5.34)$$

5.4 Design Recommendations

To properly install a bolt or rivet a minimum distance from the center of the fastener to any edge of the member must be maintained. A minimum L/d ratio of 1.5 is suggested as this conforms to current practice.

The design region shown in Fig. 5.52 is further bounded by a vertical line at a bearing stress-tensile strength ratio of 1.5. This prevents bearing stresses which may lead to excessive hole deformations and the upsetting of material in front of the fastener. Although the strength in such a situation is still adequate, large deformations may limit usefulness. Furthermore, a high σ_b/σ_u^P ratio corresponds to a large ratio of bolt diameter to the plate thickness. Thin plates which may deform out of their plane due to instability of the end section may limit the ultimate capacity of the end zone. These conditions may arise if the lap plates of a butt joint are critical in bearing. Due to "catenary action," the end of the lap plates tend to bend outward. A high compressive force on the end panel may cause a dishing type failure and decrease the ultimate bearing strength.

DESIGN RECOMMENDATIONS BEARING STRESSES

Allowable Stress Design

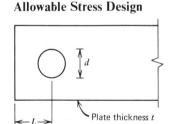

Bearing Stress $\sigma_b = P/dt$
σ_u^P = tensile strength plate material

Following conditions are to be satisfied:

(1) $L/d \geq 0.5 + 1.43\, \sigma_b/\sigma_u^P$; alternately $L/d \geq 2\sigma_b/\sigma_u^P$
(2) $L/d \geq 1.5$
(3) $\sigma_b/\sigma_u^P \leq 1.5$

b. Load Factor Design. A lower bound to the shear resistance of the end zone behind the fastener was expressed as (see Section 5.2.9):

$$F = (2t)\left(L - \frac{d}{2}\right)(0.7\,\sigma_u^P) \tag{5.35}$$

A Φ factor of 0.85 is believed adequate to account for the uncertainties in the strength of the end zone. Hence the shear strength of the end zone panel for load factor design becomes

$$\Phi F = (0.85)(1.4)\left(L - \frac{d}{2}\right) t\sigma_u^P \qquad (5.36)$$

A minimum L/d ratio equal to 1.5 is desired for installation. To limit deformations of the hole the bearing ratio σ_b/σ_u^P should not exceed 3.0 at the factored load level.

A Φ factor of 0.85 provides bearing stresses on the fastener that are equal to those obtained by factoring the allowable bearing stress values given by Eq. 5.33.

DESIGN RECOMMENDATIONS BEARING STRESSES

Load Factor Design

Shear strength end zone

$$F = (1.4)\left(L - \frac{d}{2}\right) t\sigma_u^P$$

Reduction factor Φ = 0.85
Following conditions are to be satisfied

(1) Design load × load factor ≤ Φ F; alternately $L/d \geq 1.7$ σ_b/σ_u^P
(2) $L/d \geq 1.5$
(3) $\sigma_b/\sigma_u^P \leq 3.0$

References

5.1 W. H. Laub and J. R. Phillips, *The Effect of Fastener Material and Fastener Tension on the Allowable Bearing Stresses of Structural Joints*, Report 243.2, Fritz Engineering Laboratory, Lehigh University, Bethlehem, June 1954.

5.2 R. A. Hechtman, T. R. Flint, and P. L. Koepsell, *Fifth Progress Report on Slip of Structural Steel Double Lap Joints Assembled with High Tensile Steel Bolts*, Department of Civil Engineering, University of Washington, Seattle, February 1955.

References

5.3 R. A. Hechtman, D. R. Young, A. G. Chin, and E. R. Savikko, "Slip Joints Under Static Loads," *Transactions ASCE*, Vol. 120, 1955, pp. 1335–1352.

5.4 A. A. van Douwen, J. de Back, and L. P. Bouwman, *Connections with High Strength Bolts*, Report 6-59-9-VB-3, Stevin Laboratory, Department of Civil Engineering, Delft University of Technology, Delft, the Netherlands, 1959.

5.5 O. Steinhardt and K. Möhler, *Versuche zur Anwendung Vorgespannter Schrauben im Stahlbau, Teil II*, Bericht des Deutschen Ausschusses für Stahlbau, Stahlbau-Verlag Gmbh, Cologne, Germany, 1959.

5.6 G. H. Sterling, and J. W. Fisher, "A440 Steel Joints Connected by A490 Bolts," *Journal of the Structural Division, ASCE*, Vol. 92, ST3, June 1966.

5.7 J. R. Divine, E. Chesson, Jr., and W. H. Munse, *Static and Dynamic Properties of Bolted Galvanized Structures*, Department of Civil Engineering, University of Illinois, April 1966.

5.8 A. Kuperus, *The Ratio Between the Slip Factor of Fe 52 and Fe 37, C.E.A.C.M. X-6-27*, Stevin Laboratory, Department of Civil Engineering, Delft University of Technology, Delft, the Netherlands, 1966.

5.9 G. C. Brookhart, I. H. Siddiqi, and D. D. Vasarhelyi, *The Effect of Galvanizing and Other Surface Treatment on High Tensile Bolts and Bolted Joints*, Department of Civil Enginerring, University of Washington, Seattle, September 1966.

5.10 J. H. Lee and J. W. Fisher, *The Effect of Rectangular and Circular Fillers on the Behavior of Bolted Joints*, Report 318.4, Fritz Engineering Laboratory, Lehigh University, Bethlehem, Pa., June 1968.

5.11 J. H. Lee, C. O'Connor, and J. W. Fisher, "Effect of Surface Coatings and Exposure on Slip," *Journal of the Structural Division, ASCE*, Vol. 95, ST11, November 1969.

5.12 G. L. Kulak and J. W. Fisher, "A514 Steel Joints Fastened by A490 Bolts," *Journal of the Structural Division, ASCE*, Vol. 94, ST10, October 1968.

5.13 J. R. Divine, E. Chesson, Jr., and W. H. Munse, *Static and Dynamic Properties of Bolted Galvanized Structures*, Department of Civil Engineering, University of Illinois, Urbana, April 1966.

5.14 C. C. Chen and D. D. Vasarhelyi, *Bolted Joints with Main Plates of Different Thicknesses*, Department of Civil Engineering, University of Washington, Seattle, January 1965.

5.15 D. D. Vasarhelyi and K. C. Chiang, "Coefficient of Friction in Joints of Various Steels," *Journal of the Structural Division, ASCE*, Vol. 93, ST4, August 1967.

5.16 U. C. Vasishth, Z. A. Lu, and D. D. Vasarhelyi, "Effects of Fabrication Techniques," *Transactions ASCE*, Vol. 126, 1961, pp. 764–796.

5.17 M. Maseide and A. Selberg, *High Strength Bolts used in Structural Connections*, Division of Steel Structures, Technical University of Norway, Trondheim, Norway, January 1967.

5.18 K. Klöppel, and T. Seeger. *Sicherheit und Bemessung Von H. V. Verbindungen Aus ST37 und ST52 Nach Versuchen unter Dauerbelastung und Ruhender Belastung*, Technische Hochschule, Darmstadt, Germany, 1965.

5.19 N. G. Hansen, "Fatigue Tests of Joints of High Strength Steels," *Journal of the Structural Division, ASCE*, Vol. 85, ST3, March 1959.

5.20 P. C. Birkemoe, D. F. Meinheit, and W. H. Munse, "Fatigue of A514 Steel in Bolted Connections," *Journal of the Structural Division, ASCE*, Vol. 95, ST10, October 1969.

5.21 J. W. Fisher, and J. L. Rumpf, "Analysis of Bolted Butt Joints," *Journal of the Structural Division, ASCE*, Vol. 91, ST5, October 1965.

5.22 J. W. Fisher, "Behavior of Fasteners and Plates with Holes," *Journal of the Structural Division, ASCE*, Vol. 91, ST6, December 1965.

5.23 G. L. Kulak, "The Analysis of Constructional Alloy Steel Bolted Plate Splices," Ph.D. Dissertation, Lehigh University, Bethlehem, Pa., June 1967.

5.24 R. Kormanik and J. W. Fisher "Bearing Type Bolted Hybrid Joints," *Journal of the Structural Division, ASCE*, Vol. 93, No. ST5, October 1967.

5.25 J. W. Fisher and G. K. Kulak, "Tests of Bolted Butt Splices," *Journal of the Structural Division, ASCE*, Vol. 94, ST11, November 1968.

5.26 V. H. Cochrane, "Rules for Rivet Hole Deduction in Tension Members," *Engineering News Record*, Vol. 80, November 16, 1922.

5.27 W. G. Brady and D. C. Drucker, "Investigation and Limit Analysis of Net Area in Tension," *Transactions ASCE*, Vol. 120, 1955, pp. 1133–1154.

5.28 W. H. Munse and E. Chesson, "Riveted and Bolted Joints: Net Section Design," *Journal of the Structural Division, ASCE*, Vol. 89, ST1, Part 1, February 1963.

5.29 E. Chesson and W. H. Munse, "Riveted and Bolted Joints: Truss-Type Tensile Connection," *Journal of the Structural Division, ASCE*, Vol. 89, ST1, Part 1, February 1963.

5.30 European Convention for Constructional Steelwork, *European Guidelines on the Use of High Strength Bolts in Steel Constructions*, Document CECM-X-70-8D, 3rd ed. Rotterdam, The Netherlands, April 1971.

5.31 W. H. Munse, *The Effect of Bearing Pressure on the Static Strength of Riveted Connections*, Bulletin 454, Engineering Experiment Station, University of Illinois, Urbana, July 1959.

5.32 J. Jones, "Bearing-Ratio Effect on Strength of Riveted Joints," *Transactions ASCE*, Vol. 123, 1958, pp. 964–972.

5.33 L. A. Aroian, "The Probability Function of the Product of Two Normally Distributed Variables," *Annals of Mathematical Statics*, Vol. 18, p. 265, 1947.

5.34 K. C. Chiang and D. D. Vasarhelyi, *The Coefficient of Friction in Bolted Joints Made with Various Steels and with Multiple Contact Surfaces*, Department of Civil Engineering, University of Washington, Seattle, February 1964.

5.35 R. A. Bendigo, R. M. Hansen, and J. L. Rumpf, *A Pilot Investigation of the Feasibility of Obtaining High Bolt tensions Using Calibrated Impact Wrenches*, Fritz Lab Report 200.59. 166A, Lehigh University, Bethlehem, Pa., November 1959.

5.36 J. de Back and L. P. Bouwman, *The Friction Factor Under Influence of Different Tightening Methods of the Bolts and of Different Conditions of the Contact Surfaces*, Stevin Laboratory, Report 6-59-9-VB-3, Delft University of Technology, Delft, The Netherlands, August 1959.

5.37 S. Hojarczyk, J. Kasinski, and T. Nawrot, "Load Slip Characteristics of High Strength Bolted Structural Joints Protected from Corrosion by Various Sprayed Coatings," *Proceedings, Jubilee Symposium on High Strength Bolts*, the Institution of Structural Engineers, London, 1959.

5.38 G. L. Kulak, "The Behavior of A514 Steel Tension Members," *Engineering Journal AISC*, Vol. 8, No. 1, January 1971.

5.39 J. de Back and A. de Jong, *Measurements on Connections with High Strength Bolts, Particularly in View of the Permissable Arithmetical Bearing Stress*, Report 6-68-3, Stevin Laboratory, Delft University of Technology, Delft, the Netherlands, 1968.

5.40 M. Hirano, "Bearing Stresses in Bolted Joints," *Society of Steel Construction of Japan*, Vol. 6, No. 58, Tokyo, 1970.

5.41 K. L. Johnson and J. J. O'Connor, "Mechanics of Fretting," *Proceedings of the Institution of Mechanical Engineers*, Vol. 178, Part 3J, London 1963-1964.

5.42 P. C. Birkemoe and R. S. Srinivasan, "Fatigue of Bolted High Strength Structural Steel," *Journal of the Structural Division, ASCE*, Vol. 97, No. ST3, March 1971.

5.43 T. R. Gurney, "The Effect of Mean Stress and Material Yield Stress on Fatigue Crack Propagation in Steel," *Metal Construction and British Welding Journal*, Vol. 1, No. 2, February 1969.

5.44 J. Tajima and K. Tomonaga, *Fatigue Tests on High-Strength Bolted Joints*, Structural Design Office, Japanese National Railways, Tokyo, 1963.

5.45 E. Chesson, Jr., "Bolted Bridge Behavior During Erection and Service," *Journal of the Structural Division, ASCE*, Vol. 91, ST3, June 1965.

5.46 A. Nadai, *Theory of Flow and Fracture of Solids*, Vol. 1, 2nd ed., McGraw-Hill, New York, 1950, p. 229.

5.47 I. Fernlund, *A Method to Calculate the Pressure Between Bolted or Riveted Plates*, Report 17, Inst. Machine Elements, Chalmers University of Technology, Gothenburg, 1961.

5.48 J. W. Carter, K. H. Lenzen, and L. T. Wyly, "Fatigue in Riveted and Bolted Single-Lap Joints," *Transactions ASCE*, Vol. 120, 1955.

5.49 J. W. Fisher and L. S. Beedle, "Criteria for Designing Bearing-Type Bolted Joints," *Journal of the Structural Division, ASCE*, Vol. 91, ST5, October 1965.

5.50 G. S. Vincent, *Tentative Criteria for Load Factor Design of Steel Highway Bridges*, American Iron and Steel Institute, February 1968.

5.51 J. W. Fisher, P. Albrecht, B. T. Yen, D. J. Klingerman, and B. M. McNamee, *Fatigue Strength of Welded Beams*, NCHRP Report 147, Highway Research Board National Academy of Sciences, 1974.

Chapter Six
Truss-Type Connections

6.1 INTRODUCTION

Chapter 5 summarized the strength, behavior, and design of flat plate joints where the shear planes of the joint components were all parallel to one another. Many structural connections combine plates and rolled shapes of various configurations. Axially loaded members in truss systems are typical examples. Shapes such as angles, tee sections, and wide flange sections have been used for the members in truss systems with a small to moderate span length. For larger truss systems, two or more shapes may be combined to form a "built-up" member. Some of the commonly used types of built-up sections are shown in Fig. 6.1. The solid lines indicate the sections that extend the full length of the member, and the dashed lines show the nonload carrying components that are used to maintain the geometry of the member.

In some welded truss systems the adjoining members are often directly welded together. In bolted and riveted trusses, gusset plates are generally used to transfer the load from one member to another. Depending on member geometry, single or double gusset lap joints may be used as indicated in Fig. 6.2. Past experience with these types of connections has yielded satisfactory behavior when the joint is subjected to static loading conditions. However, riveted hangers and truss members have exhibited fatigue crack growth in the connection region of members which has often resulted in failure of the member. Because of these failures, the stress patterns and load transfer in this type of connection have been extensively studied.[5.28, 5.29, 6.1-6.4] Although valuable information was obtained from these studies, the static and fatigue strength for these types of connections continue to be difficult to predict. The test results do provide some guidance on the more significant factors. In this chapter emphasis is placed on factors which affect the connected member. The design and behavior of the gusset plate is discussed in Chapter 15.

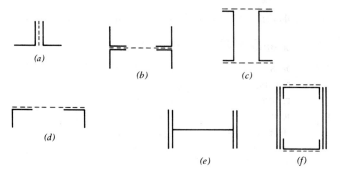

Fig. 6.1. Typical built-up truss members.

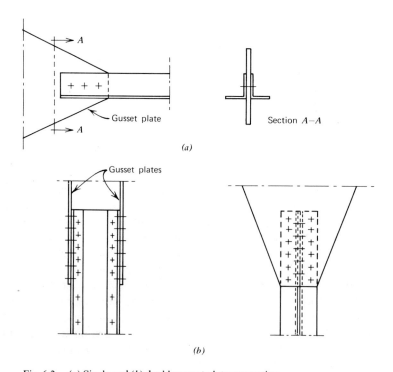

Fig. 6.2. (a) Single and (b) double gusset plate connection.

6.2 BEHAVIOR OF TRUSS-TYPE CONNECTIONS

6.2.1 Static Loading

Built-up members in truss systems (see Fig. 6.1) are subjected to either compression or tension forces. Unless buckling governs, tension-type loading is more critical than compressive loading as far as strength of the member and the connection is concerned. The reduced area of the net section of tension members yields the most critical cross-section. The force in the member must be transferred to gusset or splice plates by shear and bearing of the fasteners, or by friction between the faying surfaces of slip-resistant joints. Similar load transfer conditions exist in butt splices. Unlike plate butt splices, the centroidal axis of member components of built-up sections are not as close to each other, and this generally results in a load transfer which differs significantly from the load transfer observed in plate butt joints. The resulting eccentricity of forces affects the member's strength and behavior.

Tests have indicated that the net section efficiency of built-up members shows a significant variance, mainly as a result of joint geometry in the connection region.[5.28, 5.29, 6.3] One of the major factors that influences the effectiveness of the net section is the distribution of the cross-sectional material relative to the gusset plates or other elements that connect the member at panel points. All member components are assumed to be uniformly stressed at some distance from the connection region. Measurements have shown this to be a reasonable assumption.[6.4] A nonuniform stress distribution is created in the connection region, because not all member components are connected to the gusset plates. For instance, the load carried by the outstanding leg of the angles shown in Fig. 6.2a is transferred through fasteners placed in the other leg of the angle. Similarly, the load carried by the web of the member shown in Fig. 6.2b is transferred to the gusset through the fasteners placed in the flange angles. Generally this results in higher stresses in the connected components which are attached to the gusset plates. Depending on joint geometry and material characteristics, this may result in a decrease in efficiency of the net section in the connection region as these components tend to reach ultimate strength before the complete net section capacity has been developed. Similar results were observed in tests of angles welded to a gusset plate.[6.5] In such situations care should be taken that the actual section strength will be sufficient to cause failure of the member by yielding of the member outside the connection (see Section 5.2.6). This loss of efficiency as a result of the distribution of cross section material relative to the gusset plate is often referred to as "shear lag."

Fig. 6.3. Angle failure in built-up section. (Courtesy of University of Illinois.)

Munse and Chesson have examined the behavior of various cross sections. They observed that the loss in efficiency at the net section due to shear lag was related to the ratio of the length L of the connection and the eccentricity $\bar{x}$ from the face of the gusset plate to the center of gravity of the connected component (see Fig. 6.4a).[5.28, 5.29] To determine $\bar{x}$ for a symmetric cross-section that is connected to two gusset plates (see Fig. 6.4b), the member should be considered as two parts symmetrical about the longitudinal axis. The parameter $\bar{x}/L$ accounts for the effectiveness of the cross-section material with respect to the shear plane between the member and the gusset plate. The significance of this factor is discussed hereafter.

The unequal distribution of fastener loads in a butt joint was discussed in Chapter 5. A similar load distribution occurs among the fasteners of a built-up section. Hence relatively high loads are transferred by the end fasteners. As a result, fastener failures similar to the ones observed in long symmetric butt joints have been observed in built-up members as well.[5.28, 5.29, 6.3]

6.2 Behavior of Truss-Type Connections

The length L of the connection not only affects the load distribution among the fasteners but also influences the shear lag in a connection. Munse and Chesson concluded that a decrease in joint length increases the shear lag effect. This conclusion was based on test results from connections of the type as shown in Fig. 6.2b, which were tested to failure with either five or 10 A325 bolts in the connection region.[5.29] In both cases failure of the members occurred in the net section at the first line of fasteners, as illustrated in Fig. 6.3. The member with five bolts in a line had less strength (about 18%) at the net section compared to the longer joint with 10 bolts in a line. The fasteners were not the critical components for either test joint. Since the geometry of both joints was the same except for the joint length, it was concluded that the efficiency of the net section increases with a decrease in the ratio of $\bar{x}/L$.[5.28, 5.29] Hence an increase in joint length generally increases the effectiveness of the net section, but decreases the effectiveness of the fasteners.

To approximate the efficiency of the net section by taking into account joint length and joint geometry, Munse and Chesson suggested that the actual net area be reduced to an effective net section area by applying a reduction factor V to account for the shear lag.[5.28, 5.29] The reduction factor V was defined by the following empirical relationship

$$V = 1 - \left(\frac{\bar{x}}{L}\right) \tag{6.1}$$

where L is the joint length and $\bar{x}$ is the eccentricity between the shear plane and the centroidal axis of the connected component (see Fig. 6.4). Hence the effective net section area of a built-up member is equal to

$$\text{effective area} = A_n\left(1 - \frac{\bar{x}}{L}\right) \tag{6.2}$$

where A_n is the net area of the connected member.

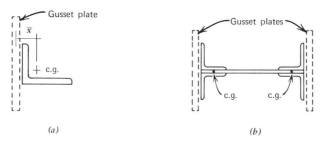

Fig. 6.4. Schematic of eccentricity in built-up joints.

Although shear lag is the major factor that reduces the efficiency of the net section, other factors such as ductility of the material, the ratio of the fastener gauge g to the fastener diameter d, and fabrication procedures also influence the efficiency of the net section. Figure 5.26 showed that the A_n/A_g ratio influences the tensile strength of the material of planar tension specimens. Generally an increase in tensile strength accompanied a decrease in A_n/A_g.

Built-up sections are often connected to the gusset plates only through relative small portions of the total cross-sectional area. Only these portions are subject to an increase in strength because of their A_n/A_g ratio. Therefore, the influence of the A_n/A_g ratio on the net section of the member is less pronounced than in butt-type joints.

Ductility of the member material affects the net section strength as well as the load distribution among the fasteners. An increase in ductility tends to increase the net section strength and provides a more uniform load transfer among the fasteners.

It was pointed out in Section 2.7 that punched holes should be reamed to remove the work-hardened material which exhibits low ductility and may

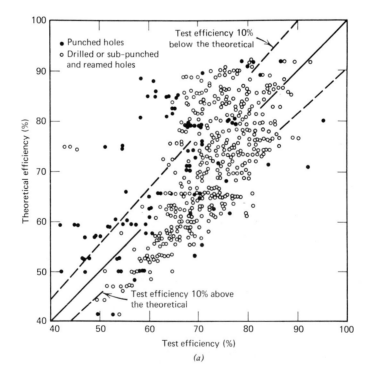

(a)

6.2 Behavior of Truss-Type Connections

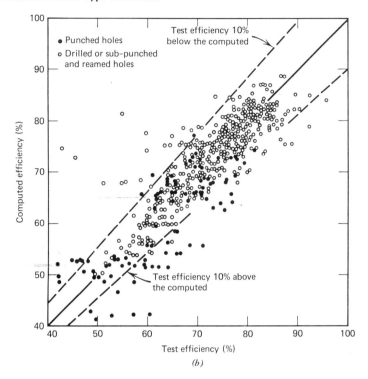

Fig. 6.5. Correlation of theoretical and test efficiencies. (*a*) based on net area; (*b*) based on Eq. 6.2.

contain small cracks as a result of the fabrication process. For these reasons, joints with punched holes often show a decreased efficiency when compared to similar sections with punched and reamed holes or drilled holes. This condition can be more critical if substantial shear lag exists as well.[5.28]

Munse and Chesson developed empirical relationships to account for these factors mentioned above. They compared the observed efficiency of test data with the efficiency of a member computed on the basis of the net section without accounting for the influence of factors such as shear lag, and such. As expected, a significant scatter of the data resulted, as shown in Fig. 6.5*a*. The scatter of data reduced when the observed test efficiency was compared with a computed efficiency that accounted for such factors as shear lag, ductility of the material, fastener spacing, and fabrication procedure. This is illustrated in Fig. 6.5*b*. They concluded that for most of the connections examined, shear lag was the major factor causing the difference between experimental and predicted efficiencies. The correlation between experimental and predicted efficiencies was further improved when the ductility of the material and the fabrication procedures were taken into account.[5.28, 5.29]

As indicated in Fig. 6.1e, rolled or welded built-up H-shapes are commonly connected by gusset plates attached to the flanges. The eccentricity, $\bar{x}$, is the distance between the shear plane and the centroidal axis of the connected component. The connected component in this instance is equivalent to a T-section. The web shear lag condition has been observed in moment connections with flange splice plate.

6.2.2 Repeated Loading

The fatigue strength of built-up structural shapes, especially in the connection region, has been the concern of many engineers as experience has shown this to be a critical factor for repeatedly loaded structures. Several failures of riveted members in truss bridges, constructed of built-up sections, were attributed to fatigue.[6.4] A detailed analysis of all the factors involved is not possible, but some guidance can be obtained from an examination of these failures.

A survey of the fatigue failures observed in riveted bridges showed that the fatigue cracks in built-up members often initiated from the side of a rivet hole at the edge of the gusset plate or splice plates (see Fig. 6.6). When cracks occurred in the gusset plate, they started at the sides of the rivet holes at the end of the members, as indicated in Fig. 6.6. Severe stress concentrations provided by geometry and shear lag in combination with the initial flaw conditions at those points made those locations susceptible to crack growth. The initial flaw condition for these joints is basically not

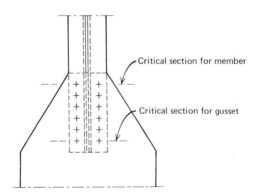

Fig. 6.6. Critical sections for a joint in a built-up section subjected to fatigue loading conditions.

different from conditions encountered in other bolted or riveted splices. Small microcracks at the sides of the hole are present as a result of the fabrication process. The stress concentration in connections of built-up sections is likely to be more severe than encountered in symmetric butt splices because of the resulting eccentricities and shear lag. This is more severe in riveted joints, because the clamping force is not as great as in bolted joints and more localized bearing occurs. Stress concentrations at the end rivet holes are further aggravated by the unequal load distribution among the fasteners. Sometimes these conditions may not significantly influence the static strength of the connection. However, the fatigue strength is adversely affected.

The fatigue strength is improved when rivets are replaced by high-strength bolts. This procedure has been used to overcome fatigue related problems in existing riveted bridge joints. The high clamping force in the bolt results in a much better stress condition at the critical sections at the fasteners holes. If sufficient slip resistance is provided, bearing stresses are eliminated and crack initiation, and growth is not as critical at fastener holes.

Because of symmetry and the existence of a web plate the connections shown in Fig. 6.2 do not develop severe secondary stresses from out-of-plane bending. When eccentrically loaded members are used and these secondary deformations are not prevented by proper lacing or diaphragms, the member tends to align and this results in additional bending stresses. Although the static strength is not greatly affected,[6.2, 6.5] severe reductions in fatigue strengths have been observed.[6.2] Net section as well as gross section fatigue failures developed prematurely in eccentrically loaded members and depended on the loading and the joint geometry. Reductions in life up to 80% were observed when compared to data obtained from tests on similar symmetric butt splices.[6.2] This reduction is due to severe stress conditions caused by the secondary stresses resulting from out-of-plane deformations. These tests indicated clearly the need for proper restraints of the connection if the possibility of fatigue failure is to be minimized. When restraints to out-of-plane bending are provided, the fatigue strength of bolted connections in built-up truss members is comparable to the fatigue strength of similar butt joints.

6.3 DESIGN RECOMMENDATIONS

The design recommendations given in Section 5.4 for bolts in slip-resistant and bearing-type joints are applicable to the design of connections in built-up members. Although the load distribution among the fasteners in joints for built-up sections is not identical to plate butt splices, the difference is considered negligible for practical purposes.

The static strength of the net section of a tension member was shown to be affected by several factors.[5.28, 5.29] For design the influence of shear lag seems dominant over the other factors and hence should be considered in design. The effective net section should reflect the influence of shear lag. The empirical formula (Eq. 6.2), proposed by Chesson and Munse,[5.28, 5.29] provides a reasonable estimate of the effective net section area. The effective net area $\bar{A}_n$ is provided by

$$\bar{A}_n = A_n \left(1 - \frac{\bar{x}}{L}\right) \tag{6.3}$$

To achieve yielding on the gross section of the member before the tensile strength of the net section is reached, the design recommendations for the A_n/A_g ratio, as given by Eq. 5.4.7 should be satisfied. To account for shear lag, the net section area A_n is to be replaced by the effective net section area $\bar{A}_n$ of the member.

Present AASHO specifications incorporate shear lag effects in tension members consisting of single angles or T-sections by assuming the effective net section area to be equal to the net area of the connected leg or flange and one-half of the area of the outstanding leg.[2.2] Additional requirements regarding the effective net section are provided for some other joint geometries. These requirements have greater applicability when members are subjected to cyclic loading.

When fatigue is to be considered in the design of a joint or net area for a built-up section, sufficient restraints should be provided to prevent secondary stresses from developing. Slip-resistant joints are preferred for high fatigue strength. The design recommendations given in Chapter 5.4 for butt-type joints are applicable to these types of joints when secondary stresses are minimized. The governing net-section stress should be evaluated on the basis of an effective net section, to account for the stress raising effects due to shear lag and other factors.

References

6.1 AREA Committee on Iron and Steel Structures, "Stress Distribution in Bridge Frames—Floorbeam Hangers," *Proceedings, American Railway Engineering Association*, Vol. 51, 1950, pp. 470–503.

6.2 K. Klöppel and T. Seeger, "Dauerversuche Mit Einschnittigen HV-Verbindugnen Aus ST37," *Der Stahlbau*, Vol. 33, No. 8, August, and No. 11, October 1964.

6.3 E. Chesson, Jr., and W. H. Munse, "Behavior of Riveted Truss Type Connections," *Transactions, ASCE*, Vol. 123, 1958, pp. 1087–1128.

References

6.4 L. T. Wyly, M. B. Scott, L. B. McCammon, and C. W. Lindner, *A Study of the Behavior of Floorbeam Hangers*, American Railway Engineering Association Bulletin 482, September, October 1949.

6.5 G. J. Gibson and B. T. Wake, "An Investigation of Welded Connections for Angle Tension Members," *Journal of the American Welding Society*, Vol. 7, No. 1, January 1942.

Chapter Seven

Shingle Joints

7.1 INTRODUCTION

In contrast with butt-type splices, the main components of the members of shingle joints are spliced at various locations along the joint. By terminating the main plates at different locations, the continuation plate can also serve as a cover plate over several regions of the joint (see Fig. 7.1). This type of connection provides a more gradual transfer of load in the plates throughout the joint. The connection is often used where the main member consists of several plies of material. Typical examples are the built-up box sections of chord members of truss bridges.

Shingle joints result in less joint thickness than butt joints, since the butt joint requires all the force to be transferred into the lap plates. In a shingle joint the load is carried by the lap plates as well as the continuous main plates at each plate discontinuity. Shingle joints can also facilitate the connection of various bridge components in a truss bridge. For example, plate A in Fig. 7.1 may also serve as a gusset for other members framing into the chord.

Shingle joints are most often used where reversal of stress is unlikely to occur because of the large dead load. Hence most shingle joints are not slip critical, and joint strength, rather than slip, is the governing criteria. Because special situations may require a design to be slip-resistant, design recommendations for both types of load transfer are given.

7.2 BEHAVIOR OF SHINGLE JOINTS

Figure 7.2 shows a typical load-deformation curve for a shingle joint.[7.1] This particular joint consisted of three regions with six $7/8$-in. A325 bolts in each region. The plates had clean mill scale surface condition and the yield strength of the plate material was about 50 ksi. The load-deformation curve shown in Fig. 7.2 indicates that in the early load stages the load is completely carried by the frictional forces acting on the faying surfaces. Tests have demonstrated that shingle joints often exhibit two distinct load levels at which major slip occurs. At the first slip load, movement develops

7.2 Behavior of Shingle Joints

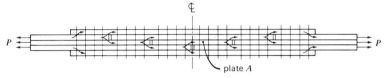

Fig. 7.1. Force flow in typical triple plate shingle joint.

mainly along the shear plane adjacent to the main plate terminations. This slip plane is depicted as plane A in Fig. 7.2. At first, little slip or no movement was observed along the second slip plane, indicated as plane B in Fig. 7.2. Upon increasing the load, a second major slip occurs with slip developing along the second slip plane (plane B in Fig. 7.2). At the same time some additional slip develops along the first slip plane (plane A).

It has been observed in tests on shingle joints that the total amount of slip tends to be less than the hole clearance.[7.1, 7.5] This is especially true for large and complex bolted joints mainly because of unavoidable misalignment tolerances during the fabrication process.

After major slip, the behavior of shingle joints is in many respects similar to the behavior of symmetric butt joints. Since the fasteners are bearing

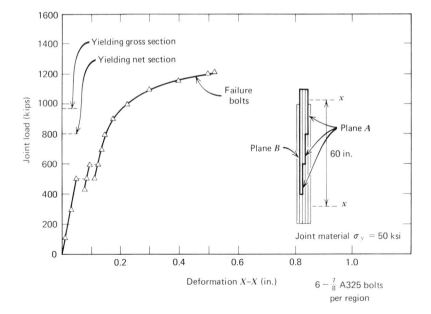

Fig. 7.2. Load-deformation behavior of shingle joint.

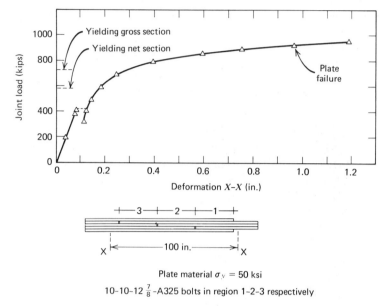

Fig. 7.3. Load-deformation behavior of shingle joint.

against the plate material, fastener deformations are developed in proportion to the load transmitted by each fastener. At high load levels the load deformation relationship of the joint becomes nonlinear because of plastic deformations in the fasteners and the plates. Depending on the joint geometry and the mechanical properties of the constituent parts, failure occurs either by shearing of the fasteners or by fracture of the plates. Both types of failures have been experienced in tests.[7.1, 7.5] Characteristic load deformation curves are shown in Figs. 7.2 and 7.3.

Although both shingle and symmetric butt joints yield similar load deformation relationships, the deformation pattern of the individual fasteners is usually quite different. This is illustrated in Fig. 7.4 where a sawed section of a three-region joint is shown after the joint was tested to failure.* The end fastener has sheared off, and it is visually apparent that the bolt deformation decreased rapidly from the end fastener toward the middle of the joint. An apparent double shear condition existed in the first six or

---

* To use the same bolt lot in all tests it was necessary (see Fig. 7.4) for the bolts in this particular joint to have less than full thread engagement for the nuts. Control tests indicated that the full bolt shear capacity was obtained even with less than full thread engagement. This practice, however, is not recommended for field installations.

7.2 Behavior of Shingle Joints

seven fasteners of region 1, as indicated by the deformation along both shear planes. Thereafter, the fasteners resisted the load in single shear, transferring the load primarily to the lap plates adjacent to the main plate cutoffs. In a symmetric butt joint the fasteners are loaded in double shear, whereas the fasteners in a shingle joint may be loaded either in single or double shear, depending on their location within the joint.

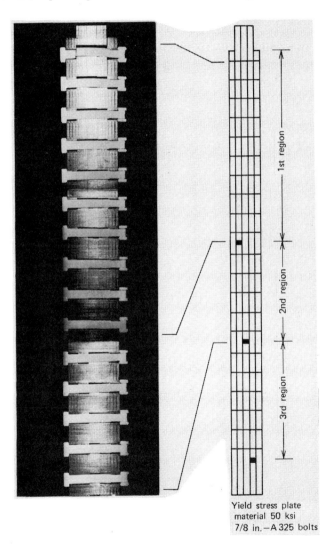

Yield stress plate
material 50 ksi
7/8 in.—A 325 bolts

Fig. 7.4. Sawed section of a three-region shingle joint after loading to failure.

Tests on riveted shingle joints yielded an overall behavior which was comparable to the behavior of bolted shingle joints.[3.8, 7.3] Riveted joints exhibited less slip than the bolted joints, because there is less hole clearance. When fastener failure is the governing failure mode, the overall deformation of large riveted shingle joints is likely to exceed the comparable deformation of an identical bolted joint.[7.2] This is primarily because of the different load-deformation characteristics of rivets as compared to high-strength bolts.

7.3 JOINT STIFFNESS

The stiffness of a joint is characterized by the slope of its load-deformation diagram. Figures 7.2 and 7.3 indicate that initially the total load is transferred by friction on the faying surfaces of the joint. It is also apparent that the stiffness of shingle joints is not significantly affected by slip of the connection. Only yielding of the gross or net section causes a decrease in joint stiffness. Since the working load level does not exceed the yield strength of the net section, the joint stiffness may be considered equivalent to the full cross-section with an area equal to the total gross area of the main and lap plates. A comparable condition was observed with symmetric butt joints.

7.4 LOAD PARTITION AND ULTIMATE STRENGTH

The analytical solution for load partition and ultimate strength of shingle joints is based on a mathematical model, which is similar to the symmetric butt joints described earlier. The butt joint is a special case of a shingle joint.[7.2] The same basic assumptions which are discussed in Section 5.2.5 still apply. In addition, it is assumed that the transfer of load between the lap plates and the main plate takes place along the two planes that are common to the main plate core as illustrated in Fig. 7.5. Thus no relative movement between the various plies of the lap plate or between the various plies of the main plate is considered. Each segment of the lap plate and main plate between consecutive fasteners is assumed to function as a unit

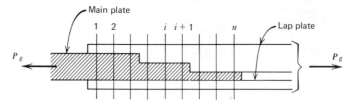

Fig. 7.5. Idealized model of a shingle joint.

7.5 Effect of Joint Geometry

with properties which are aggregate of the constituent plies. The model assumes the top and bottom lap plates to be a single plate of variable thickness, comparable to the main plate. This idealization results in regions of variable length with uniform plate properties within each region.

The force displacement relationships for plies of uniform width as well as for the fasteners are those empirically developed in Ref. 5.22. The solution is comparable to the solution for a symmetrical butt splice.[7.2] The theoretical results were in good agreement with the experimental data on bolted shingle joints.[7.1] It was concluded that the load partition and ultimate strength can be predicted within acceptable limits, if double shear behavior is assumed in the first region and single shear behavior in the interior regions of shingle joints. This assumption is examined in greater detail in Section 7.5.

7.5 EFFECT OF JOINT GEOMETRY

The theoretical solution was used to study analytically the effects of various joint geometries on the ultimate strength.[7.1] The nondimensionalized ratio of the predicted ultimate strength to the working load of the joint, P_u/P_w, was used as an index of joint behavior. The working load was either based on the fastener shear area or the net area of the main plate. Two possible assumptions for evaluating the total fastener shear in a joint were examined, namely (a) double shear of the fasteners throughout the joint, and (b) double shear in the first region and single shear in the other regions.

In the analytical study the yield stress and tensile strength of the plate material were assumed as 60 and 88 ksi respectively, resulting in a 35-ksi allowable tensile stress for the plate material. The joints were fastened by $7/8$-in. A325 bolts of minimum specified mechanical properties. The fastener pitch was held constant at 3 in.

The variables studies were (a) the A_n/A_s ratio, defined as the ratio of the net main plate area in the first region to the total effective fastener shear area; (b) the total number of fasteners in a joint; (c) the number of fasteners per region; and (d) the number of regions.

7.5.1 Effect of Variation in A_n/A_s Ratio and Joint Length

Figure 7.6 shows the change in joint strength with length for different A_n/A_s ratios ranging from 0.375 to 1.00 for shingle joints with three equal length regions. The fasteners were assumed to act in double shear in all three regions for one series of studies, and the results are indicated by the open dots. Each curve represents a different allowable shear stress, for example an A_n/A_s ratio of 0.625 corresponds to an allowable shear stress

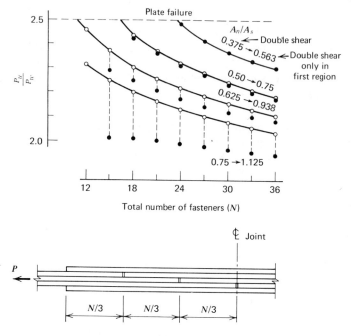

Fig. 7.6. Effect of assuming single shear in interior regions. ○ Analytical prediction assuming double shear. ● Analytical prediction assuming double shear in region 1; single shear in interior regions.

of 22 ksi for double shear. Test results have indicated that the joint strength is likely to be overestimated for joints with high A_n/A_s ratios. This was primarily due to the single shear behavior observed in the interior regions.[7.1, 7.3]

The analysis was also made assuming single shear behavior of the fasteners in the interior regions. The results are summarized in Fig. 7.6. It is apparent that for lower A_n/A_s ratios it does not matter whether double or single shear is assumed in the interior regions. For these joints the fasteners in the first region are the critical ones, as is illustrated in Fig. 7.7. At higher A_n/A_s levels the load carried by the interior fasteners was greater, and a reduction in effective shear area had a more pronounced influence on joint strength (see Fig. 7.6). This was confirmed by the experimental results.[7.1]

7.5.2 Number of Fasteners per Region

The effect of varying the number of fasteners in each region was studied analytically by shifting an equal number of fasteners from each interior

7.5 Effect of Joint Geometry

region into the first region. The total number of fasteners in the joint as well as the plate areas were maintained. Double shear behavior of the fasteners was assumed in the first region and single shear behavior in the interior regions. The results are summarized in Fig. 7.8. Sometimes a fastener failure was predicted in the interior regions when the fasteners were rearranged.[7.1] At the 0.75 A_n/A_s level, this only occurred in the short joints when four fasteners were shifted into the first region. No variation in strength occurred in the longer joints.

At the 1.125 A_n/A_s level, slight increases in strength were predicted by shifting fasteners into the first region.

From this study it was concluded that the predicted strength of shingle joints of a given length was not greatly influenced by rearranging the fasteners. This trend was also confirmed by the test data reported in Ref. 7.1.

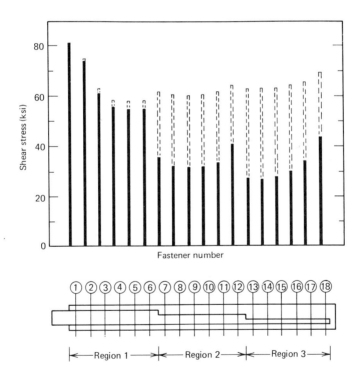

Fig. 7.7. Fastener shear distribution assuming single or double shear in interior regions A_n/A_s = 0.500. Ultimate load 930 kips. Double shear assumed in all regions. A_n/A_s = 0.750. Ultimate load 927 kips. Double shear in first region. Single shear in interior regions.

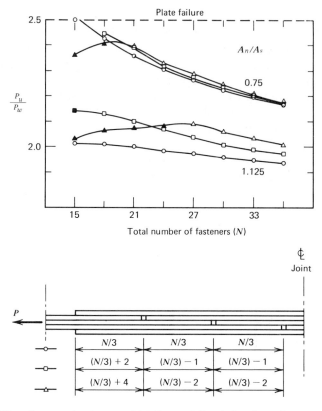

Fig. 7.8. Effect of rearranging fasteners. ■ ▲ Denote failure in interior regions.

7.5.3 Number of Regions

The effect of varying the number of main plate terminations was studied by comparing the strengths of joints with one, two, and three regions. All joints had the same total number of fasteners as well as the same plate areas. Also, multiple region joints provided an equal number of fasteners per region. Double shear behavior of the fasteners was assumed in the first region with single shear in the interior regions. The one-region joints were symmetrical butt joints having the total main plate area terminated at one location.

Figure 7.9 shows the change in ratio P_u/P_w due to the variation in the number of regions. Note that the A_n/A_s ratio increases as the number of regions increases. This results from the assumed shear behavior of the fasteners in the interior regions. As indicated in Fig. 7.9 for the joints repre-

7.6 Design Recommendations

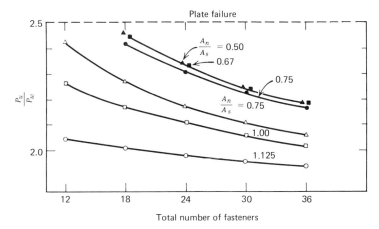

Fig. 7.9. Effect of number of regions. ▲ △ One-region joint (double shear is assumed). ■ □ Two-region joint: double shear in first region and single shear in interior regions. ● ○ Three-region joint; $A_n/A_s = 0.50$ open symbols for the single region joint; $A_n/A_s = 0.75$ solid symbols for the single region joint.

sented by the solid dots (A_n/A_s ratio is equal to 0.50 for the single region joint), there was no appreciable change in strength as the number of regions was changed. At the higher A_n/A_s ratios, indicated by the open dots in Fig. 7.9, the two- and three-region joints were less efficient. Greater variation was apparent for the shorter lengths. However, it is doubtful that short joints will be shingled.

At higher A_n/A_s ratios, the distribution of load to the interior fasteners was greater than at lower A_n/A_s ratios. Thus terminating the main plates at different locations and reducing the effective shear area resulted in a reduction in strength.

7.6 DESIGN RECOMMENDATIONS

7.6.1 Approximate Method of Analysis

Shingle joints like other types of connections are statically indeterminant; thus, the distribution of forces depends on the relative deformations of the component members and fasteners. The condition is further complicated in shingle joints by the unsymmetric positioning of main plate terminations. Analytical elastic solutions that predict the distribution of load in the main and splice plates of shingle joints have been developed.[7.5] The solution has been extended into the plastic range so as to predict the ultimate strength

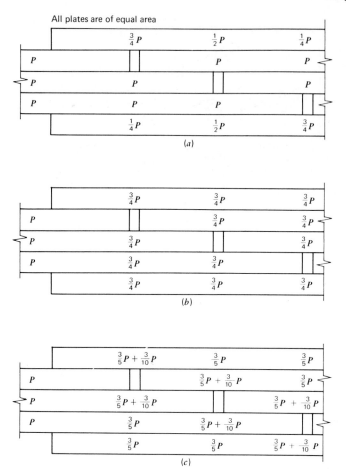

Fig. 7.10. Illustration of design methods. (*a*) Method 1; (*b*) method 2; (*c*) method 3.

of the connection.[7.2] These theoretical analyses, however, are too cumbersome and impractical for ordinary design practice. Simplifying assumptions must be made that reduce the solution for design to one based primarily on equilibrium.

There are several existing methods for estimating the distribution of force in the main and lap plates of a shingle splice. Two of the most popular methods are:[7.4]

1. Forces in splice plates are inversely proportional to their distances from the member being spliced.

7.6 Design Recommendations

2. Forces in each member at a section through a splice are proportional to their areas.

In method 1, it is assumed at each discontinuity that the amount of force distributed to the lap plates is proportional to the area of the member being terminated. The forces in the continuous main members are assumed to remain unchanged. This is illustrated schematically in Fig. 7.10a. The transfer of load is made in the region directly preceding the point of termination, and it is assumed that the original load is restored to the spliced member in the region following the termination.

In method 2 (see Fig. 7.10b), the total applied load is assumed to be distributed to all continuous members at the position of a main plate termination in proportion to their areas. No direct assumption is made regarding the amount of load transferred to the splice plates in a particular region as in method 1. If the lap plates are of equal area, method 2 predicts that the shear transfer is equal along the top and bottom shear planes in the first region regardless of their positions with respect to the member being terminated.

Previous shingle joint tests have shown that at each plate discontinuity, there was a sudden pick-up of load in the adjacent plate elements.[3.8, 7.5] Another approximate method of analysis was developed on the basis of these observations and test results. This method, referred to as method 3, and illustrated in Fig. 7.10c, assumes that the total load is distributed to all members at a section through the joint in proportion to their areas, first considering the terminated members as being continuous. The load assumed to be carried by a terminating member is then distributed to the two adjacent plates in proportion to their areas. Hence a two stage distribution is used.

Figure 7.11 compares the measured plate forces in a three-region test joint with the three design methods.[7.1] The partition of load was determined from the measured plate strains at different cross sections along the length. The comparisons were at the working load levels as determined by the main plate net areas. It is apparent from Fig. 7.11 that method 1 underestimated the total transfer of load in the first and second region. Loads substantially greater than estimated by method 1 were measured in the bottom lap plates. Test results indicated that the force in the top and bottom plates were nearly equal in the first region.

The actual distribution of load in the main plates of the joint determined by method 2 were in good agreement with the measured forces. Slight variation between the theoretical distribution and test results occurred in the top and bottom lap plates. It was found that this method slightly underestimates the forces in the plates adjacent to a plate termination.

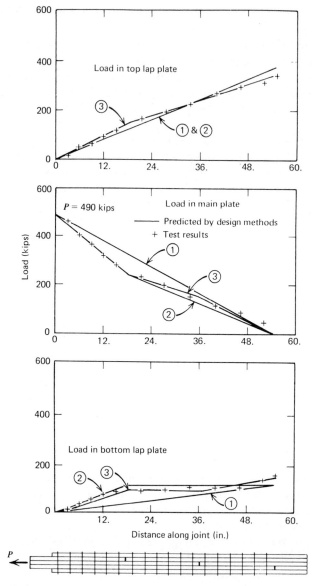

Fig. 7.11. Comparison of design methods with test results.

7.6 Design Recommendations

The distributions of force determined by method 3 provided the best correlation with the test results, as shown in Figs. 7.7 and 7.8. The method provided a reasonable estimate of the force distributions in all joint components, and accurately predicts a more effective use of the fasteners in the interior regions, thus requiring less fasteners than the other methods; this method is therefore recommended for design purposes.

For design it is recommended that method 3 be used to approximate the load distribution in the plates and fasteners. With this method, it is also recommended that the first region of shingle splices have double lap plates of equal area. This reduces the critical shear transfer along the plane adjacent to the first plate termination.

Where practical, it is also recommended that the top and bottom lap plates have equal lengths in the first region. As shown in Fig. 7.4, equal deformation was observed along both shear planes at failure. It is believed that equal length splice plates would more effectively utilize the critical end fasteners.

With the introduction of a gusset into the splice as in a truss joint, however, additional fasteners are required along the shear plane adjacent to the gusset to transfer load from diagonal members. Since these fasteners are not required along the bottom shear plane, it is believed that the bottom lap plates can be shorter than the top lap plate in the first region if an adequate number of fasteners is still provided.

7.6.2 Connected Material

Once the load distribution throughout the plates is determined, the plate dimensions can be obtained. The design recommendations given in Section 5.4.3 for the connected plates are applicable to shingle joints.

7.6.3 Fasteners

After the load partition has been established, the required number of fasteners per region can be determined. The difference in plate load between two adjacent plates is transmitted by shear of the fasteners. An examination of all possible shear planes in each region results in one or more critical shear planes for each region. The number of fasteners is readily determined from the shear resistance of the fasteners.

The design recommendations given in Section 5.4.2 for slip-resistant and other bolted joints subjected to static loading conditions are also applicable to the design of slip-resistant and other bolted shingle joints. The design shear stress for shingle joints depends on the bolt quality as well as on the joint length. Since the first region is the critical one in most shingle joints, the design shear stress for non-slip-critical shingle joints should be reduced by 20% if the length of the first region exceeds 50 in. All other design recommendations given in Section 5.4.2 are applicable to shingle joints.

References

7.1 E. Power and J. W. Fisher, "Behavior and Design of Shingle Joints," *Journal of the Structural Division, ASCE*, Vol. 98, No. ST9, September 1972.

7.2 S. Desai and J. W. Fisher, *Analysis of Shingle Joints*, to be published.

7.3 E. Davis, G. B. Woodruff, and H. E. Davis, "Tension Tests of Large Riveted Joints," *Transactions, ASCE*, Vol. 66, No. 8, Part 2, pp. 1193–1299, 1940.

7.4 W. J. Yusavage (Ed.), *Simple Span Deck Truss Bridge*, Manual of Bridge Design Practice, 2nd ed., State of California, Highway Transportation Agency, Department of Public Works, Division of Highways, Sacramento, 1963.

7.5 U. Rivera and J. W. Fisher, *Load Partition and Ultimate Strength of Shingle Joints*, Fritz Laboratory Report 340.6, Lehigh University, Bethlehem, Pa., 1970.

Chapter Eight
Lap Joints

8.1 INTRODUCTION

As contrasted to symmetric butt splices, fasteners in lap splices have only one shear plane. Depending on the geometry of the joint and the loading conditions, the behavior of lap joints may differ significantly from the behavior of symmetric butt joints with the fasteners loaded in double shear.

The simplest type of lap splice is shown in Fig. 8.1a. Such joints are simple to fabricate and erect but are often avoided because of concern with their inherent eccentricity which can result in deformations as shown in Fig. 8.1a. These effects of bending may be minimized by providing restraining diaphragms or stiffeners that restrict the rotation and out-of-plane displacement of the joint. Such restraints may be an integral part of the member. Often situations arise in which the restraints are provided by the connected members itself; a typical example is the hanger connection shown in Fig. 8.1b or the flange splices of a girder (Fig. 8.1c). Because of symmetry of the shearing planes and diaphragm action of the web, bending of the lap splice does not occur in significant amount, although the fasteners are in a single shear condition and an eccentricity of the load exists.

Fasteners in a lap splice are mainly subjected to axial shear conditions. However, depending on joint geometry and loading conditions, bending can result in an additional tensile component in the fastener. As noted in the following sections, this tensile component is often of minor importance and does not affect significantly the ultimate strength of the connection.

8.2 BEHAVIOR OF LAP JOINTS

In a discussion of the behavior of lap joints it is convenient to define two categories of lap joints as follows:

1. Joints in which restraints are provided so that bending can be neglected (Fig. 8.1b and c).
2. Joints that are not restrained against bending. In these joints secondary bending stresses are developed due to the eccentricity of the load.

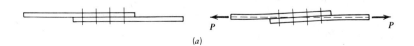
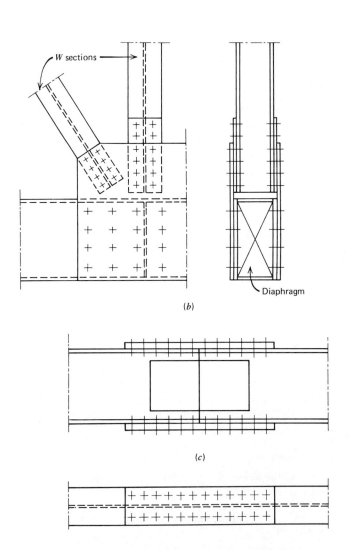

Fig. 8.1. Typical lap splices with fasteners subjected to single shear. (*a*) Lap splice connection; (*b*) typical connection in truss-type bridge; (*c*) girder-splice.

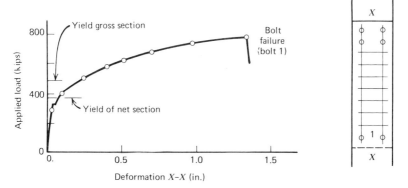

Fig. 8.2. Typical load-deformation curve for lap joints in which restraints against bending are provided.

Fig. 8.3. Single-shear specimen after test. (Courtesy of U.S. Steel Corp.)

Static tension tests of lap joints with restraint against out-of-plane deformation exhibit a load deformation behavior which is essentially comparable to the behavior observed for symmetric butt joints (see Fig. 8.2). The slip resistance and the ultimate strength of single shear lap splices was found to equal one-half the double shear resistance provided by a butt joint. As expected, the "unbuttoning" behavior (as discussed in Chapter 5) was also observed in long lap joints.[4.6, 8.1]

The load-deformation behavior of lap joints which were not restrained against out-of-plane displacement has been examined with small joints with two or three fasteners in a line.[6.2, 8.2, 8.3] Since restraints were not provided, the joints showed considerable deformation due to the eccentricity of the load, as shown in Fig. 8.3. It is evident that the effects of bending are mainly confined to the regions where plate discontinuities occur. Obviously as the joint length increases, bending will become less pronounced. The influence on the behavior of the connection should decrease. The influence of bending is most pronounced in a splice with only a single fastener in the direction of the applied load. In such a joint the fastener is not only subjected to single shear, but a secondary tensile component may be present as well. Furthermore, the plate material in the direct vicinity of the splice is subjected to high bending stresses due to the eccentricity of the load. However, this has little influence on the load capacity, as the material will strain harden and cause yielding on the gross area of the connected plate.

Tests on single bolt lap splices showed that the slip resistance was not noticeably affected by the additional bending.[8.2, 8.3] Shear failures of the fasteners were observed at an average fastener shear stress which was about 10% less than observed in symmetric butt joints with similar material properties. Hence the bending tended to decrease slightly the ultimate strength of short connections. The shear strength of longer lap joints with no restraints against bending is believed to be not as affected by the effects of bending.

Lap joints may be subjected to a repeated type loading as well. The critical joint component under such loading conditions is not the fastener but the plate material. A severe decrease in the plate fatigue strength is apparent in unrestrained lap joints when compared to butt joints.[6.2] The bending deformations cause larger stress ranges to occur at the discontinuities of the joint. The bending stress combines with the normal stress and results in high local stresses that reduce the fatigue strength. The reduction in fatigue strength depends on the joint geometry and the magnitude of the secondary bending. Hence single shear splices subject to stress cycles should not be used unless the out-of-plane bending deformations are prevented.[6.2]

8.3 DESIGN RECOMMENDATIONS

When designing lap joints, both the fasteners and the plate material should be considered. Consideration should also be given to the type of loading and whether out-of-plane deformation will adversely affect the joint performance.

8.3.1 Static Loading Conditions

It was concluded before that the average shear strength of the fasteners at ultimate load and the slip resistance of lap joints are in reasonable agreement with the behavior observed on comparable symmetric butt joints. Therefore, the design recommendations given in Chapter 5 are applicable to lap joints for static type loading conditions. Bending of the joint does not significantly influence the slip resistance or strength. Hence the provisions for bolts and plate material are applicable.

8.3.2 Repeated Type Loading

Since the plate is the critical element under repeated loads, lap joints should only be used under repeated loading conditions when secondary bending stresses are prevented or minimized. This requires suitable stiffening or joint geometry, which will prevent out-of-plane movement. Lap connections that are susceptible to out-of-plane movements should not be used under repeated loading conditions. The design recommendations given in Chapter 5 for the plate material of symmetric butt joints are applicable as well to the design of lap joints which are not subjected to bending effects.

References

8.1   R. A. Bendigo, J. W. Fisher, and J. L. Rumpf, *Static Tension Tests of Bolted Lap Joints*, Fritz Engineering Laboratory Report 271.9, Lehigh University, Department of Civil Engineering, Bethlehem, Pa., August 1962.

8.2   Z. Shoukry and W. T. Haisch, "Bolted Connections with Varied Hole Diameters," *Journal of the Structural Division, ASCE*, Vol. 96, ST6, June 1970.

8.3   K. D. Ives, *Evaluation of Oversize Holes in Friction-Type Single Shear Joints*, Bulletin Applied Research Laboratory, U.S. Steel Corporation, Pittsburgh, Pa., June 1971.

Chapter Nine
Oversize and Slotted Holes

9.1 INTRODUCTION

Since the first application of high-strength bolts in 1947, bolt holes $\frac{1}{16}$ in. larger than the bolts have been used for assembly. A similar practice was adopted in Europe and Japan, where a hole diameter 2 mm greater than the nominal bolt diameter became standard practice.[9.1]

Restricting the nominal hole diameter to $\frac{1}{16}$ in. in excess of the nominal bolt diameter can impose rigid alignment conditions between structural members, particularly large joints. Sometimes erection problems occur when the holes in the plate material do not line up properly because of mismatching. Occasionally, steel fabricators must preassemble structures to ensure that the joint will align properly during erection. With a larger hole size, it is possible to eliminate the preassembly process and save both time and money. To determine the feasibility of oversize holes, it was necessary to evaluate the performance of bolted connections with greater amounts of oversize.

An oversize hole provides the same clearance in all directions to meet tolerances during erection. If, however, an adjustment is needed in a particular direction, slotted holes can be used as shown in Fig. 9.1a and b. Depending on the direction of the slots with respect to the direction of the applied load, slotted holes are identified by their parallel or transverse alignment (see Fig. 9.1a and b).

Oversize and slotted holes result in additional plate material removed from the vicinity of high clamping forces. The influence of this condition on the behavior of connections has been investigated experimentally.[4.26, 8.2, 8.3, 9.1] The effect of oversize and slotted holes on such factors as the loss in bolt tension after installation, the slip resistance and the ultimate strength of shear splices has been examined. Tightening procedures were studied as well. Provisions based on these findings are now included in specifications.[1.4]

9.2 EFFECT OF HOLE SIZE ON BOLT TENSION AND INSTALLATION

The load-deformation characteristics of joints assembled with high-strength bolts installed in oversize or slotted holes depend, among other factors, on

9.2 Effect of Hole Size on Bolt Tension and Installation

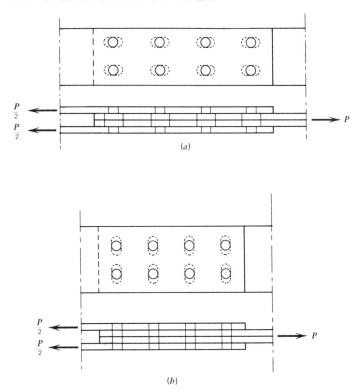

Fig. 9.1. Slotted holes. (a) Parallel slotted holes; (b) transverse slotted holes.

the bolt clamping force. Hence it is necessary to examine the effect of varying hole diameters on the bolt installation. This includes the degree of scouring and the clamping force induced by standard installation procedures. These factors are of primary interest when slip-resistant joints are used.

Tests have indicated that oversize and slotted holes can influence significantly the level of bolt preload when bolts are installed in accordance with common practice.[4.26] This is illustrated in Fig. 9.2 where the observed bolt tension after installation by the turn-of-the-nut method is shown for several hole clearances.[4.26] One-inch-diameter A325 bolts were installed in joints with $\frac{1}{4}$-in. hole clearance with and without washers under the turned element. The average bolt tensions for both types of joints with $1\frac{1}{4}$-in.-diameter holes were about equal. The achieved bolt tension averaged 118% of the minimum required tension which was about 15% lower than the average tension that resulted in joints with $\frac{1}{16}$-in. hole clearance. Plate depressions

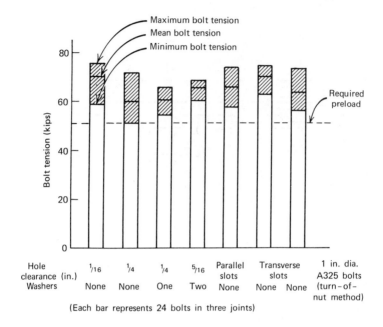

Fig. 9.2. Range of bolt tensions.

occurred under the bolt heads during tightening and were greater than the depressions observed with the usual $\frac{1}{16}$-in. hole clearance. Severe galling of both plate and nut occurred with oversize holes when washers were omitted from under the turned element as is illustrated in Figs. 9.3 and 9.4.[4.26] One-inch-diameter bolts installed with only one washer under the turned element in $1\frac{5}{16}$-in.-diameter holes failed to achieve their minimum required tension. The bolt heads had recessed severely into the plate around the holes. Therefore, washers were placed under the nut and bolt head. The range of bolt tension achieved with washers under nut and bolt head ranged from 110 to 144% of the minimum required tension, with an average value of 125%.

The difficulty in achieving the minimum required tension results from the bolt depression into the plate around the hole. The rotation of the nut does not result in the degree of bolt elongation desired. When a washer was not used under both head and nut, installing 1-in.-diameter bolts in a $1\frac{5}{16}$-in. hole produced an excessive bearing pressure between the bolt head and the plates, resulting in severe indentations. After one-half turn of the nut from snug position, the minimum required bolt tension was not achieved. Bolts that were installed in holes with $\frac{1}{4}$-in. clearance did achieve the

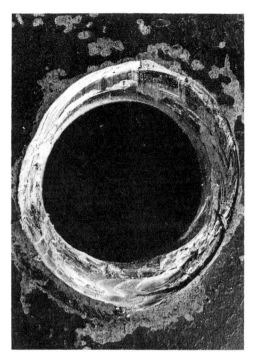

Fig. 9.3. Severe galling of plate under turned element-(1/4 in. clearance, no washer).

Table 9.1. Hole Clearance for Different Hole Sizes

Bolt Size	Maximum Hole Diameter (in.)	Amount Clearance
$\frac{1}{2}$	$\frac{11}{16}$	$\frac{3}{16}$
$\frac{5}{8}$	$\frac{13}{16}$	$\frac{3}{16}$
$\frac{3}{4}$	$\frac{15}{16}$	$\frac{3}{16}$
$\frac{7}{8}$	$1\frac{1}{16}$	$\frac{3}{16}$
1	$1\frac{1}{4}$	$\frac{1}{4}$
$1\frac{1}{8}$	$1\frac{7}{16}$	$\frac{5}{16}$
$1\frac{1}{4}$	$1\frac{9}{16}$	$\frac{5}{16}$
$1\frac{3}{8}$	$1\frac{11}{16}$	$\frac{5}{16}$
$1\frac{1}{2}$	$1\frac{13}{16}$	$\frac{5}{16}$

Fig. 9.4. Plate area under element in which washer was used (1/4 in. clearance).

minimum required bolt tension. Assuming that the bearing pressure developed under the flat areas of the bolt heads with $\frac{1}{4}$-in. clearance holes is the maximum permitted on A36 steel plate, the maximum hole clearance for any size bolt can be determined. The area of the plate remaining under the flat of the bolt head must be sufficient so that this pressure is not exceeded. The results of such computations are summarized in Table 9.1. All of the hole diameters have been rounded off to the nearest sixteenth of an inch.

Bolts installed by the turn-of-nut method in slotted holes also showed a decrease in the mean bolt tension when compared to similar bolts installed in standard holes with a $\frac{1}{16}$ in. oversize.[4.26] Hence the use of oversize or slotted holes is likely to reduce slightly the mean clamping force in the fastener.

Immediately after a bolt is tightened, a loss in bolt tension occurs. This is thought to result from creep and plastic deformation in the threaded portions and plastic flow in the steel plates under the head and the nut. These deformations result in an elastic recovery and loss in bolt tension.

9.3 Joint Behavior

Studies on bolts installed in holes with a standard hole clearance are summarized in Ref. 4.26 and in Chapter 4. In general, the total loss in preload was about 5 to 10% of the initial preload, depending on grip length (3 to 6 in.) and whether washers were used. Most of the loss in preload occurred within a short time after the bolt was tightened.

A few relaxation tests have been conducted on bolts installed in oversize holes and are reported in Ref. 4.26. It was observed that none of the variations in the hole diameter or the presence of slots had any significant effect on this loss. Virtually all of the losses occurred within 1 week after installation as was the case with earlier studies. The loss in tension was observed to be about 8% of the initial preload which was directly comparable to earlier studies on regular size holes with a standard clearance of $\frac{1}{16}$ in.

9.3 JOINT BEHAVIOR

9.3.1 Slip Resistance

Figure 9.5 shows typical load-slip relationships of joints with oversize or slotted holes.[4.26] The load-slip response is almost linear until the load approaches the major slip load. The initial slip was never observed to be equal to the hole clearance of the joint. Subsequent loading of the joint after major slip was initiated, produced small slips until the joint came into bearing. These small slips occurred at loads near the major slip load.

Tests were performed on double shear splices with 1-in.-diameter A325 bolts, as shown in Fig. 9.1.[4.26] A summary of the observed slip coefficients as a function of the hole geometry is shown in Fig. 9.6. It was concluded that the average slip coefficient for joints with up to $\frac{1}{4}$-in.-hole clearance did not change with varying oversize. The joints with $\frac{5}{16}$-in.-clearance holes showed a 17% decrease in the slip coefficient for clean mill scale faying surfaces. The slip coefficient for joints with slotted holes showed a 22 to 33% decrease when compared to test specimens with a hole clearance of $\frac{1}{16}$ in. A decrease in slip resistance with the removal of plate material from around the bolt was expected because of the resulting high-contact pressures in the area around the bolt. Removal of the plate causes extremely high-contact pressures adjacent to the bolt holes which tends to flatten the surface irregularities. This reduces the slip resistance of the joint.

The slip resistance is also affected by the decreased clamping force which has been observed in joints with oversize and slotted holes. The combined effects of the change in slip coefficient and the reduction in the clamping force on the slip resistance is estimated to cause a 15% reduction in slip resistance for oversize holes and a 30% reduction for parallel and transverse slotted holes.[4.26]

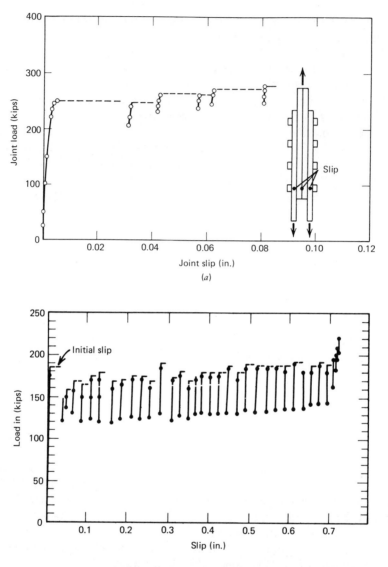

Fig. 9.5. Typical load slip diagrams. (a) Load slip diagram of joint with oversize holes. (b) Load slip diagram of joint with slotted holes.

9.4 Design Recommendations

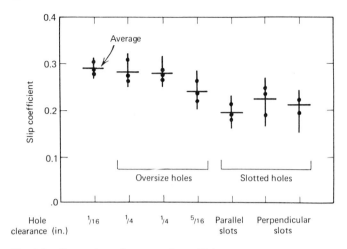

Fig. 9.6. Comparison of average slip coefficients.

Major slip of the connection is terminated when one or more bolts come into bearing against the plates. The amount of slip exhibited before bearing occurs depends on the available clearance and fabrication tolerances. Joints with oversize holes or parallel slotted holes may undergo substantial displacements if the slip resistance of the joint is exceeded.

9.3.2 Ultimate strength

The ultimate strength of a connection is governed by either the shear capacity of the bolts or the tensile capacity of the plates. Tests have shown that transverse slotted holes do not reduce the tensile strength of the net area of the plates or the shear strength of the bolts.[4.26] Hence the ultimate strength of a joint is not affected by either oversize or slotted holes.

9.4 DESIGN RECOMMENDATIONS

Since the ultimate strength of a joint with oversize or slotted holes is the same as the ultimate strength of a similar standard type connection with identical bolt and plate areas, the design recommendations given in Chapter 5 are applicable. The provisions for both plate material and bolts of bearing-type shear splices are applicable to joints with oversize or slotted holes. Care must be exercised when using oversize or slotted holes to ensure that excessive deformation will not occur at working loads. The slots should be oriented so that large displacements cannot result. Transverse slotted holes are preferable, since they limit the slip to the same magnitude that can be experienced with standard hole clearances.

Design recommendations for slip-resistant joints with oversize or slotted holes must reflect the reduced slip resistance. Hole diameters that do not exceed those given in Table 9.1 do not significantly alter the slip coefficient. However, the clamping force is reduced by about 15% and must be reflected in the slip resistance and design conditions. A factor 0.85 provides for the reduced clamping force and its effect on the slip resistance. For slip-resistant joints with slotted holes, a reduction factor of 0.70 accounts for the loss in slip resistance caused by either parallel or slotted holes.

To prevent the use of extremely large slotted holes, present specifications limit the length of slotted holes to $2\frac{1}{2}$ times the bolt diameter (these are defined as long slotted holes). The width of the hole should not exceed the bolt diameter by more than $\frac{1}{16}$ in. Short slotted holes are also used. Short slotted holes are $\frac{1}{16}$ in. wider than the bolt diameter and have a length that does not exceed the allowable oversize diameter for that bolt size by more than $\frac{1}{16}$ in. Joints with short slotted holes will develop the same slip resistance as joints with oversize holes. Therefore, the design of joints with oversized or short slotted holes is the same.

To achieve an adequate clamping force in the bolts, washers should be used under the bolt head and nut when oversize or slotted holes occur in the outside plates of a joint.

DESIGN RECOMMENDATIONS OVERSIZE AND SLOTTED HOLES

Hardened washers are to be inserted under both the head and the nut if oversize or slotted holes are placed in the outside plies of a connection.

Table 9.2. Oversize and Slotted Hole Dimensions

Bolt Diameter	Hole Sizes		
	Oversize Holes (Max.)	Short Slotted Holes (Max.)	Long Slotted Holes (Max.)
$\frac{5}{8}$	$\frac{13}{16}$	$\frac{11}{16} \times \frac{7}{8}$	$1\frac{11}{16} \times 1\frac{9}{16}$
$\frac{3}{4}$	$\frac{15}{16}$	$\frac{13}{16} \times 1$	$\frac{13}{16} \times 1\frac{7}{8}$
$\frac{7}{8}$	$1\frac{1}{16}$	$\frac{15}{16} \times 1\frac{1}{8}$	$\frac{15}{16} \times 2\frac{3}{16}$
1	$1\frac{1}{4}$	$1\frac{1}{16} \times 1\frac{5}{16}$	$1\frac{1}{16} \times 2\frac{1}{2}$
$1\frac{1}{8}$	$1\frac{7}{16}$	$1\frac{3}{16} \times 1\frac{1}{2}$	$1\frac{3}{16} \times 2\frac{13}{16}$
$1\frac{1}{4}$	$1\frac{9}{16}$	$1\frac{5}{16} \times 1\frac{5}{8}$	$1\frac{5}{16} \times 3\frac{1}{8}$
$1\frac{3}{8}$	$1\frac{11}{16}$	$1\frac{7}{16} \times 1\frac{3}{4}$	$1\frac{7}{16} \times 3\frac{7}{16}$
$1\frac{1}{2}$	$1\frac{13}{16}$	$1\frac{9}{16} \times 1\frac{7}{8}$	$1\frac{9}{16} \times 3\frac{3}{4}$

Slip-Resistant Joints

$$\tau_a = \beta_1 \beta_2 \beta_3 \tau_{\text{basic}}$$

for β_1, β_2 see Chapter 5.4. For oversize and short slotted holes not exceeding the dimensions given in Table 9.2

$$\beta_3 = 0.85$$

For long slotted holes not exceeding the dimensions given in Table 9.2

$$\beta_3 = 0.70$$

References

9.1 European Convention for Constructional Steelwork, *Specifications for Assembly of Structural Joints Using High Strength Bolts*, 3rd ed., Rotterdam, The Netherlands, April 1971.

9.2 O. Steinhardt, K. Möhler, and G. Valtinat, *Versuche zur Anwendung Vorgespannter Schrauben im Stahlbau, Teil IV*, Bericht des Deutschen Auschusses für Stahlbau, Stahlbau-Verlag Gmbh, Cologne, Germany, February 1969.

Chapter Ten

Filler Plates Between Surfaces

10.1 INTRODUCTION

Often splices are symmetric and consist of identical structural components on each side of the splice. The joint components share a number of common shear planes, and splice plates are required to transfer the load across the splice. Many joints connect members with different dimensions or else gaps are provided to permit ease of erection. When members with different dimensions are connected, the splice must be filled out to permit the faying surfaces of the splice plates to be in contact with both members. This minimizes eccentricities in the joint and provides better geometric conditions. Filler plates are used as packing pieces to create a common faying surface and shear plane on both sides of the splice. This minimizes the secondary stresses and eccentricities in the different joint components. The beam-girder splice with different depth members on each side of the joint, as illustrated in Fig. 10.1, is a typical example of a joint using filler plates. Filler plates are also frequently encountered in splices of axially loaded built-up members in truss bridges. Furthermore, they are used in the field to provide clearance for field assembly.

The influence of filler plates on the load transfer through a splice comprising one or more filler plates is briefly discussed in this chapter. Unfortunately there are not a great deal of experimental data available at the time of publication. A series of tests was carried out in England in 1965 on single bolt joints with $\frac{1}{8}$-in.-thick washers inserted between faying surfaces.[10.1] Tests were also reported by Lee and Fisher on four bolt joints with blast cleaned surfaces and fillers.[5.10] The filler thickness varied from $\frac{1}{16}$ to 1 in. Although the available data are rather limited, they provide an indication of the effect and behavior of joints with filler plates.

10.2 TYPES OF FILLER PLATES AND LOAD TRANSFER

Filler plates are classified as "loose" or "tight" fillers. In the case of loose fillers, the plates are solely used as packing pieces. They only provide a common shear plane on each side of the splice as shown in Fig. 10.2a.

10.2 Types of Filler Plates and Load Transfer

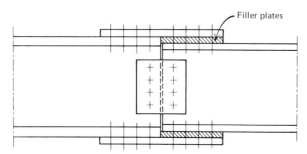

Fig. 10.1. Beam girder splice with filler plates.

Tight fillers are also used as packing pieces, but the fillers are extended beyond the splice plates or the joint is made larger. These fillers are connected by additional fasteners outside of the main splice and also provide a common shear plane. In addition, they become an integral part of the member as shown in Fig. 10.2b.

In slip-resistant joints the load is transferred by frictional forces acting on the contact surfaces. Hence the fasteners are not loaded in direct shear, as is the case in a bearing-type joint. Therefore, loose fillers are adequate for slip-resistant joints when the surface condition of the joint components provide adequate slip resistance, and the forces can all be transferred on the faying surfaces. Test results reported in Refs. 10. 1 and 5. 10 support this conclusion. The tests reported in Ref. 10.1 are summarized in Fig. 10.3. All specimens had two bolts in line, packed with $\frac{1}{8}$-in.-thick washers

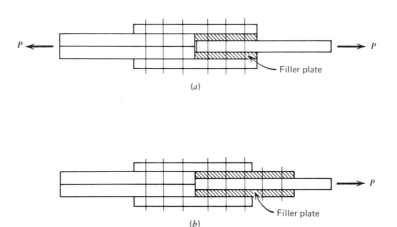

Fig. 10.2. Types of filler plates. (a) Loose fillers; (b) tight fillers.

of a variable diameter in order to control the contact area. It is readily apparent from Fig. 10.3 that the insertion of ⅛-in.-thick "loose" fillers between the joint faying surfaces did not significantly affect the slip resistance. This was observed to be true for both clean mill scale and blast-cleaned faying surfaces.

The tests reported by Lee and Fisher were on four bolt joints with blast-cleaned surfaces.[5.10] The fillers were symmetrically placed on both faying surfaces and varied in thickness from 1/16 to 1 in. Figure 10.4 shows the joints as well as some typical test results. There seems to be no significant variation in the slip resistance with different thicknesses of the fillers. Furthermore, as shown in Fig. 10.5, the observed slip coefficients varied between 0.47 and 0.57 which is within the 95% confidence limits for blast cleaned surfaces summarized in Table 5.1 from Chapter 5. It is apparent

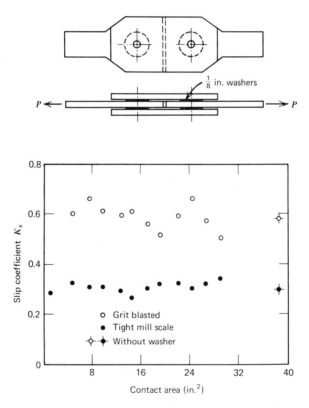

Fig. 10.3. Slip coefficient—contact area relationship for tests by Dorman Long and Company.

10.2 Types of Filler Plates and Load Transfer

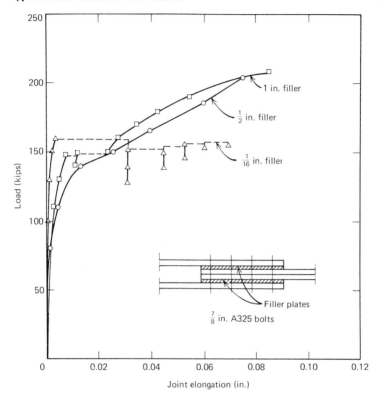

Fig. 10.4. Load versus slip behavior of joints with filler plates.

that filler plates, with a surface condition comparable to the surface condition of the main plates, do not significantly affect the slip resistance of a joint.

Vasarhelyi and Chen tested bolted butt joints with slightly different thickness main plates on each side of the joint.[10.2] Filler plates were not used and consequently full surface contact could not be obtained adjacent to the end of the thinner main plate. Generally, a decrease in slip resistance was observed when compared to the control joints with main plates of equal thickness. They suggested that the slip resistance could be improved by increasing the distance from the plate edge to the first row of bolts. This would provide more flexibility in the lap plates and allow more clamping force to be used effectively for load transfer.

Tight fillers might be advantageous or necessary if the bearing stress on the main plate rather than the shear capacity of the fastener governs the

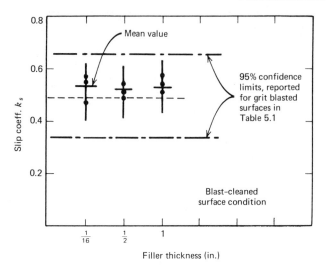

Fig. 10.5. Comparison of slip coefficients.

design. Providing a tight filler increases the thickness of the plate to be spliced. There are no bolted joints tests with tight fillers available. However, tests have been conducted on riveted joints to verify the assumed behavior.

10.3 DESIGN RECOMMENDATIONS

Depending on the required load transfer, loose or tight fillers can be used in slip-resistant or bearing-type joints. For slip-resistant joints, loose fillers with surface conditions comparable to other joint components are capable of developing the required slip resistance. Slip-resistant joints do not require additional fasteners when filler plates are used. The fillers become integral components of the joint and filler thickness does not significantly affect the joint behavior.

For bearing-type joints where the load is transmitted by shear and bearing of the bolts, loose fillers can be used as long as excessive bending of the bolts does not occur. Tight fillers are not required in bearing-type joints if the allowable bearing stress on the main plate is not exceeded. Tests on riveted joints have indicated that tight fillers are desirable when thick filler plates are needed and long grips result. This requires additional fasteners which are preferably placed outside the connection, as shown in Fig. 10.2b. As an alternative solution, the additional fasteners may be placed in the main splice.

The design recommendations given in Chapter 5 for the plates and fasteners are applicable to the design connections with filler plates.

References

10.1 L. G., Johnson, *High Strength Friction Grip Bolts*, unpublished report, Dorman Long and Company, England, September 1965.

10.2 D. D. Vasarhelyi and C. C. Chen, "Bolted Joints with Plates of Different Thickness," *Journal of the Structural Division, ASCE,* Vol. 93, ST6, December 1967.

Chapter Eleven

Alignment of Holes

11.1 INTRODUCTION

Holes in mechanically fastened joints are either punched, subpunched and reamed, or drilled. When high-strength bolts are used, the hole diameter is generally $\frac{1}{16}$ in. greater than the nominal bolt diameter. Since connections contain two or more fasteners, the alignment of holes is of concern. Shop practice usually results in separate fabrication of the constituent parts of a joint. Since dimensional tolerances are necessary during the fabrication process, the holes of component parts of a joint are not likely to be perfectly aligned. Unless all plies are clamped together before drilling, the holes may not be aligned. Misalignment may also result from erection tolerances. Hence it is desirable to ascertain whether hole offsets have detrimental effects on the joint behavior.

This chapter discusses the influence of misalignments on the behavior of high-strength bolted connections.

11.2 BEHAVIOR OF JOINTS WITH MISALIGNED HOLES

The experimental data available on joints with misaligned holes are not extensive. Vasarhelyi *et al.* have reported on a series of tests where misalignment was purposely introduced into the joint by providing mismatching holes.[11.1, 11.2]

The two major concerns with misaligned holes is whether the slip resistance is affected and whether the misalignment adversely affects the joint strength and performance. With joints transferring load by shear and bearing of the fasteners, bolts placed in misaligned holes will obviously come into bearing prior to other fasteners in the joint. If the fasteners and plates have sufficient ductility and can accomodate the unequal forces and displacements, the misalignments should not have a significant effect.

In addition to affecting the distribution of forces on the fasteners, misalignment may also influence the stress distribution in the connected plates of the joint.

Depending on the amount of misalignment in the hole pattern, tests on misaligned joints have indicated that slip generally develops more gradually

11.2 Behavior of Joints with Misaligned Holes

when compared to joints with good alignment.[11.1, 11.2] This is expected, since full hole clearance slip is prevented due to the misalignment of the holes. As slip develops, the plates come into bearing and the fasteners generally offer further resistance to the slip movement.

A series of small slips have been observed to develop at load levels considerably above the normal slip resistance.[11.1, 11.2] These partial slips bring more bolts into bearing and result in geometric self-adjustment of the joint. elements as the applied loads force alignment of the joint. The joint tends to pivot around fasteners already in bearing, and eventually this results in more bolts in bearing.

Tests have indicated that the slip resistance of a misaligned bolted joint is equal to or exceeds the slip resistance of a joint without misalignment. This is visually apparent in Fig. 11.1. As the misaligned condition was

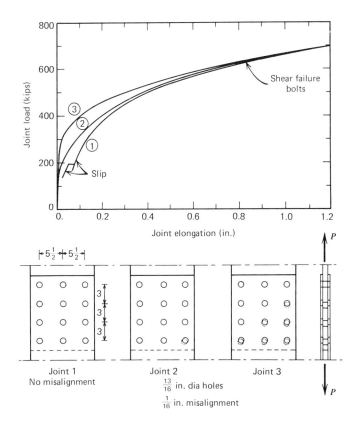

Fig. 11.1. Influence of misalignment of holes on load-deformation curve (Ref. 11.2).

made more severe, there was not as much rigid body motion possible. No significant change in joint stiffness was apparent until the applied loads were nearly twice as large as the load that caused major slip to develop with good alignment. Comparable results have been observed with more complex joints where misalignment is more probable.[3.8, 4.6] Misaligned holes always result in less movement between the connected plies. The joint stiffness is improved, and full hole slip is not possible.

When slip develops, one or more bolts come into bearing. As the applied load is increased, these bolts as well as the adjacent plate material must deform so that other bolts can come into bearing as well. If the deformation capacity of the plates and the bolts will permit it, all bolts may come into bearing before shear failure develops in one or more bolts. Excessive misalignment may prevent enough bolts from coming into bearing and prevent the full shear strength of the joint from being developed. This may result in a decrease in joint strength. This situation is somewhat analogous to the load partition that occurs in long bolted joints. The critical fastener may be subjected to severe deformations and result in premature failure prior to attaining the full joint strength.[4.6]

The tests on compact bolted joints with different degrees of misalignment throughout the bolt pattern that are summarized in Fig. 11.1 show that misalignment has a negligible effect on the ultimate strength of the joints. If anything, the misalignment had a beneficial effect. It improved the slip resistance, decreased the rigid body motion between connected plies, offered a stiffer joint, and did not result in a decrease in joint strength. Comparable results were reported in later tests.[11.2]

As the connected material increases in yield and tensile strength, misalignment may have a more adverse effect. Not as much ductility is available for the redistribution of the load, and the misaligned fastener could be prematurely sheared off. This condition is also more critical with higher strength bolts, since they have less deformation capacity in shear. The plastic deformation capacity of the plate material and the deformation capacity of the bolt both contribute to the adjustment that occurs in the joint. Obviously, the more deformation capacity that is available, the better the redistribution of plate and bolt forces.

11.3 DESIGN RECOMMENDATIONS

The amount of misalignment in a joint depends largely on the joint geometry as well as on fabrication tolerances and erection procedures. Since bolt holes are generally $\frac{1}{16}$ in. in excess of the nominal bolt diameter, some adjustment possibility is provided. Available test results do not indicate any adverse effect of misalignment resulting from hole clearance on slip resistance or strength of the joint.[11.1, 11.2] Hence the usual misalignment that may result from erection or fabrication tolerances does not affect the design of joints.

Since the deformation capacity of the fasteners and plate material are of prime importance in the readjustment capacity of bolted joints with misaligned holes, the degree of tolerance will decrease when higher strength materials with lower ductility are used.

References

11.1 D. D. Vasarhelyi et al., "Effects of Fabrication Techniques on Bolted Joints," *Journal of the Structural Division, ASCE,* Vol. 85, ST3, March 1959.

11.2 D. D. Vasarhelyi and W. N. Chang, "Misalignment in Bolted Joints," *Journal of the Structural Division, ASCE,* Vol. 91, ST4, August 1965.

Chapter Twelve

Surface Coatings

12.1 INTRODUCTION

Often situations arise in steel construction in which it is desirable to provide a protective coating on the members and the faying surfaces of their joints. The treatment prevents corrosion due to exposure before erection or provides a corrosion resistant layer to reduce maintenance costs during the lifetime of the structure. When the treatment is applied to prevent long term corrosion, the coating is of a permanent nature; usually metallic layers of zinc or aluminum are employed. For temporary protective purposes, a wash primer is often used that is usually removed upon assembly by grinding or by dissolving with various solvents. Other less permanent coatings such as vinyl washes and linseed oil are also used.

It has long been recognized that protective coatings alter the slip characteristics of bolted joints to varying degrees.[4.18, 12.7] Consequently, the design of slip-resistant joints with coated faying surfaces must reflect the influence of such treatments on the slip resistance.

For bearing-type joints in which the working load level is determined on the basis of joint strength, the coating is not critical, since the strength characteristics of the joint are not affected by protective treatments. Therefore, this chapter is confined to the influence of protective coatings on the load-deformation characteristics and performance of slip-resistant joints subjected to various types of loading.

In the past, galvanized members have been used mainly for special purposes such as transmission line towers, and the joints were often designed on the basis of their strength so that bearing-type connections resulted. In some structures ribbed bearing bolts were used to minimize joint slip. The use of coatings for slip-resistant joints was limited or prohibited by the specifications.[1.4, 5.30] These restrictions were the result of early research which indicated that a low frictional resistance resulted when galvanized surfaces were present.[12.7] As a result of continuing research, protective surface treatments that provide adequate slip resistance have been developed.[4.11, 4.18, 4.27, 5.11, 5.17, 5.37, 9.1, 12.1-12.3] These studies indicate that adequate

12.2 Effect of Type of Coating on Short-Duration Slip Resistance

frictional resistance of coated surfaces can be achieved, and also indicated that coated high-strength bolts, nuts, and washers could be used provided that a suitable lubricant was used on the threaded parts of the fastener (see Section 4.6). As a result of these studies, provisions were included in the RCRBSJ specifications in 1970 which permitted certain surface treatments to be used in slip-resistant joints.[1.4] These treatments and their influence upon the load-slip behavior of slip-resistant joints are discussed in this chapter.

12.2 EFFECT OF TYPE OF COATING ON SHORT-DURATION SLIP RESISTANCE

When only temporary protection of the faying surface is needed, paints are often placed on the weather-exposed surfaces. Vinyl-washes and linseed oil have also been used as substitutes for red lead and similar paints.[5.11] If a more permanent protective coating is required, a metallic layer with a high corrosion resistance must be applied to the structural element. The most commonly used protective coatings can be classified as follows:

1. Hot-dip galvanizing with or without a preassembly treatment to improve the slip resistance of the surface.
2. Metallizing with either sprayed zinc, aluminum, or a combination of both metals.
3. Zinc rich paints composed of organic or inorganic vehicles.

The effects of these coatings on the slip resistance of connections subjected to short-duration statically applied loads are discussed in this section. Other factors such as the load-deformation behavior under sustained or repeated loading conditions must be considered when they are applicable and are discussed in subsequent sections.

12.2.1 Hot-Dip Galvanizing

The hot-dip galvanizing process requires the removal of the mill scale prior to the coating application. Usually this is done by pickling the member in a bath of acid. Subsequently, the member is coated with a metallic layer by dipping it into a bath of hot metal. Iron–zinc alloys or pure zinc are generally used for this process.

Test results indicate that hot-dip galvanizing generally results in a low frictional resistance of the faying surfaces.[4.11, 4.18, 12.1, 12.13] Tests on joints with hot-dip galvanized faying surfaces have yielded slip coefficients between 0.09 and 0.36 with an average value of 0.19 (see Table 12.1).[12.13] The low slip resistance of galvanized surfaces as compared to clean mill scale surfaces is caused by the presence of the softer zinc layer which tends

Table 12.1. Slip Coefficients for Hot-Dip Galvanized Surfaces under Short-Duration Static Load

Ref.	Type of Treatment	Coating Thickness (mils)	Number of Tests	Average	Standard Deviation
4.18	1. Pickling in acid bath 2. Hot-dip galvanized	2.4–5.0	10	0.23	.023
12.1	1. Pickling in acid bath 2. Hot-dip galvanized	4.0	3	0.15	—
4.28	1. Pickling in acid bath 2. Hot-dip galvanized (tests performed on one bolt compression type specimens)	—	15	0.21	.08
12.1	1. Pickling in acid bath 2. Hot-dip galvanized	—	2	0.15	—
12.5	1. Pickling in acid bath 2. Hot-dip galvanized	3.2	—	0.20	—
	1. Sand blasted 2. Hot-dip galvanized	3.2	—	0.28	—

Note: 1 mil = 0.001 in. or 25.4 µm; a zinc coating of 1 oz./ft^2 corresponds to a coating thickness of 0.0017 in.

12.13	Summary Study Data from Various Sources	—	95	0.19	Value Min. .08 Max. .36 Estimated standard deviation .045

to act as a lubricant between the faying surfaces. Test results have also indicated that the slip coefficient decreases with an increase in coating thickness.[4.18, 12.1]

Variability in thickness of the metallic layer is inherent with the galvanizing process. Different treatment methods have also contributed to the variability observed for different test series. These factors are believed to be the major reasons for the relatively large scatter in the test data.[12.13]

The influence of the treatment method on the slip resistance of galvanized joints is illustrated by the test data summarized in Table 12.2. In

12.2 Effect of Type of Coating on Short-Duration Slip Resistance

these test series, all joint components were grit-blasted before pickling and subsequent dipping into the metal bath. Dipping time, cooling rate, and bath temperature were varied. For the plain uncoated blast-cleaned surfaces, an average slip coefficient of 0.73 resulted. The galvanized surfaces yielded average slip coefficients between 0.27 and 0.57.[12.8] The study indicated that the type of coating process can affect the slip resistance of the

Table 12.2. Influence of Pre-galvanizing Treatment on Slip Coefficient

Conditions	Series A	Series B	Series C
Surface condition	Grit blasted to white metal	Hot-dip galvanized	Hot-dip galvanized
Coating thickness (mils)	—	4.0	4.5
Coating structure	—	Fe–Zn alloys 40%; pure zinc 60%	Fe–Zn alloys 100%
Preparation	—	Grit blasting and successive pickling with HCl	Grit blasting and successive pickling with HCl
Zinc bath temperature	—	452°C	467°C
Dipping time	—	1 min	3 min
Cooling	—	High: air blowing within 20 sec after withdrawal and successive water quenching	Low: specimens kept over the bath surface for 3 min, successive water quenching

Ref.	Type of Treatment	Coating Thickness (mils)	Number of tests	Average	Standard Deviation
12.8	Grit-blasted series A	—	10	0.73	.05
	Grit-blasted hot-dip galvanized series C	4.2	10	0.57	.01
	Grit-blasted hot-dip galvanized series B	4.2	10	0.27	.03
5.17	Grit-blasted hot-dip galvanized	4.0	12	0.30	—

Table 12.3. Summary of Slip Coefficients of Hot-Dip Galvanized Surfaces (Determined from Compression-Type Specimens)

Surface Treatment	Average	Standard Deviation	Number of Tests
As received	0.21	.08	12
Weathered	0.20	.06	17
Wire-brushed	0.37	.01	6
Sand-blasted	0.44	.02	9
Shot-blasted	0.37	.10	6
Acetone-cleaned	0.32	.03	9
Phosphate-treated	0.38	.03	10
Chromate-treated	0.26	.02	6

coated surfaces. These results, as well as data reported in Ref. 5.17, show that blast cleaning the surface before hot-dip galvanizing results in an improvement of the slip resistance.[5.17, 12.5, 12.8] This results from the increased surface roughness due to the blast cleaning.

A significant improvement in the slip resistance of galvanized surfaces can be achieved by preassembly treatment of the contact surfaces. Among the treatments examined are wire brushing, sand or grit blasting, and a chemical treatment of the galvanized surfaces.[4.11] Wire brushing can be accomplished manually or with a power brush. A light blast cleaning which dulls the normal shiny appearance of the galvanized coating is generally sufficient. With either treatment it is essential to visibly alter the surface condition. However, the continuity of the coating should not be disrupted. A substantial increase in slip resistance has been observed for some of the treatments.[4.11, 4.27, 12.13]

Test results on small compression jigs are summarized in Table 12.3. Tests on larger tension connections with the same surface treatments have yielded somewhat lower slip coefficients. The results of the compression shear jigs clearly show that an improvement in slip resistance can be obtained by wire brushing or light blast cleaning the galvanized surfaces of the joints prior to assembly. Tests reported in Ref. 5.17 yielded the same trend. Hence treatment of hot-dip galvanized surfaces can result in a slip coefficient which is at least comparable to the coefficient for clean mill scale surfaces (see Fig. 12.1). Further tests are desirable to provide a better estimate of the slip coefficient for such surface conditions.

12.2 Effect of Type of Coating on Short-Duration Slip Resistance

As contrasted to clean mill scale or blast-cleaned surface conditions, a sudden slip does not usually occur in hot-dip galvanized joints. The observed slip is often gradual with increasing loads until the bolts come into bearing.

12.2.2 Metallizing

The metallizing process involves spraying a hot metal onto the surfaces of a structural element to provide corrosion resistance. Zinc and aluminum are commonly used for metallizing structural members.

The surface to be metallized should have all oil and grease removed and must be roughened by blasting. The sprayed metal will only bond adequately to cleaned and roughened surfaces. Sand, crushed slag, or chilled iron grit are commonly used for blast cleaning the surface. The coating is applied shortly after blast cleaning. Different spraying processes can be used and detail procedures are given in Ref. 12.11.

Short-duration slip tests on metallized surfaces have shown that high slip resistance can be achieved with these treatments.[4.18, 5.17, 5.37, 12.1, 12.5, 12.6] Test results from metallized joints with various coating thicknesses are summarized in Table 12.4. It is apparent that the slip coefficient is related to the coating thickness. When the coating is thick compared to the surface irregularities resulting from blast cleaning, a relatively low slip coefficient results. Very thin coatings, 0.0005 to 0.001 in. (15 to 25 μm), also result in relatively low slip coefficients. The optimum slip performance was achieved when the coating thickness was between 0.002 to 0.004 in. (50 to 100 μm).

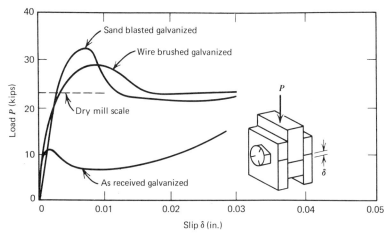

Fig. 12.1. Small shear block specimen tests indicate several surface treatments that enhance the frictional resistance of galvanized steel.

The test data also indicate a higher slip coefficient for aluminum-sprayed surfaces as compared to zinc-sprayed surfaces with the same coating thickness. This difference in behavior is believed due to the difference in hardness of the metallic layer. A higher slip coefficient results with the harder aluminum coating.

Often sealing treatments are used to improve the corrosion resistance of the surfaces and enhance their appearance.[12.11] These additional treatments tend to fill the surface irregularities and provide a smoother faying surface. This results in a decreased frictional resistance and a lower slip coefficient. Hence sealing treatments should not be used on slip resistant joints.[12.2]

Table 12.4. Slip Coefficients Metallized Surfaces (Short-Duration Tension-Type Tests)

Ref.	Type of Treatment	Coating Thickness (mils)	Number of Tests	Average	Standard Deviation
12.6	1. Corundum blast cleaned	0.8–1.6	—	0.425	—
		2.0–2.8	—	0.448	—
	2. Zinc sprayed	3.6–4.4	—	0.413	—
5.37	1. Sand blasted	8.0	—	0.400	—
	2. Zinc sprayed				
4.18	1. Sand blasted	—	2	0.480	—
	2. Zinc sprayed				
12.1	1. Grit blasted	3.0	—	0.780	—
	2. Zinc sprayed	0.6–1.0	20	0.422	.045
	1. Shot blasted	—	—	0.600	—
	2. Sprayed zinc	3.0	—	0.700	—
	1. Corundum blasted	0.6–1.0	10	0.431	.037
	2. Sprayed zinc				
12.5	1. Sand blasted	1.6	17	0.705	.049
	2. Zinc sprayed				
	1. Sand blasted	2.6	14	0.726	.049
	2. Two layers sprayed zinc				
5.17	1. Grit blasted	4.0	12	0.819	—
	2. Zinc sprayed (*compression*type specimens)				

12.2 Effect of Type of Coating on Short-Duration Slip Resistance

Table 12.4 Continued

Ref.	Type of Treatment	Coating Thickness (mils)	Number of Tests	Average	Standard Deviation
12.6	1. Corundum blasted	0.8–1.6	—	0.563	—
	2. Aluminum sprayed	2.0–2.8	—	0.575	—
		3.6–4.4	—	0.588	—
12.1	1. Shot blasted	—	—	0.640	—
	2. Aluminum sprayed	4.0	—	0.790	—
	1. Grit blasted	1.6–2.2	20	0.743	.080
	2. Aluminum sprayed	4.0	—	0.760	—
	1. Corundum blasted	1.6–2.2	10	0.728	.095
	2. Aluminum sprayed				
5.37	1. Sand blasted	10.0	—	0.400	—
	2. Aluminum sprayed				
12.5	1. Sand blasted	2.4	—	0.670	—
	2. Aluminum sprayed				
12.6	1. Corundum blasted	Layer thickness			
	2. Zinc sprayed	Zn: 1.2	—	0.490	—
	3. Aluminum sprayed	Al: 1.2			
		Zn: 1.2	—	0.420	—
		Al: 4.0			
5.37	1. Sand blasted	20.0	—	0.410	—
	2. Chrome-Nickel sprayed				
12.5	1. Sand blasted	1.6	6	0.718	.051
	2. Stainless steel sprayed				

12.2.3 Zinc-Rich Paints

Zinc-rich paints are coatings that contain a high zinc dust content and provide a hard, abrasion resistant protection for the coated surfaces.[12.12] They are mainly used for permanent or long-term corrosion protection. Some of the coatings are used for prefabrication or shop primers. The primer coats do not require as great a thickness as coatings for long-term protection.

Zinc-rich paints are available in a large number of different commercial mixes. These coatings use either organic or inorganic vehicles. Among the organic vehicles used are vinyls, epoxies, and polyesters.[5.11, 12.12] Common inorganic vehicles are silicates, phosphates, and modifications thereof. Many of these coatings are supplied with the zinc-rich pigment packaged separately, and the materials are mixed at the time of application. Depend-

Table 12.5. Slip Coefficients of Zinc-Rich Painted Surfaces (Zinc Paints and Organic Vehicles)

Ref.	Type of Treatment	Coating Thickness (mils)	Number of Tests	Average	Standard Deviation
9.2	Sand blasted	—	10	0.517	.040
	1. Sand blasted 2. Primer	0.6–0.8	8	0.203	.022
	1. Sand blasted 2. Special primer	0.8	10	0.410	.016
	1. Sand blasted 2. Zinc dust paint	0.8	10	0.392	.024
12.1	1. Sand blasted 2. Zinc dust paint	1.2	10	0.230	.030
9.2	Grit blasted	—	6	0.557	.012
	1. Grit blasted 2. One component zinc dust paint	0.6 1.2 1.8	6 6 6	0.401 0.448 0.462	.015 .012 .008
	1. Grit blasted 2. Special primer	0.6 1.2 1.8	6 6 6	0.391 0.414 0.418	.012 .021 .022
	1. Grit blasted 2. Two component zinc dust paint	0.6 1.2 1.8	6 6 6	0.299 0.309 0.328	.023 .029 .008

ing on the chemical composition, these coatings may have a pot life of as low as 6 hr.

The inorganic coatings are very resistant to solvents and oil and are also resistant to high humidity. The weathering resistance of inorganic coatings is outstanding since the coating continues to cure during prolonged exposure.[12.12] For best results, the inorganic coatings should be used over blast-cleaned surfaces that provide a "near-white" condition.

Compared with the inorganic coatings, organic coatings are generally more tolerant to variations in surface prepartion. They tend to be more flexible but are also less tough and abrasion resistant than the inorganic materials.[12.12]

The slip behavior of connections with contact surfaces treated with zinc-rich paints with inorganic or organic vehicles has been examined by

12.2 Effect of Type of Coating on Short-Duration Slip Resistance

tests.[9.2, 12.1] Table 12.5 summarizes the results of a series of tests where the faying surfaces were treated with zinc-rich paint with organic vehicles. The results of a pilot study are given in part A of Table 12.5 for surfaces which were sand blasted prior to application of the zinc paint. The paint reduced the slip coefficient to a level that was comparable with hot-dip galvanized joints.[9.2] Tests were also performed on joints treated with zinc dust paint and a special zinc based primer, both having an organic base.[9.2] The results of these tests are summarized in part B of Table 12.5. The results show that the application of a zinc-rich paint with an organic vehicle to a blast-cleaned surface will decrease the slip resistance of the blast-cleaned surfaces. It is also apparent that an increase in coating thickness from 0.0006 to 0.002 in. (15 to 50 μm) resulted in an increased slip coefficient.

The data summarized in Table 12.5 also indicate that a low slip coefficient results for two component zinc dust paints. Surface roughness measurements of the joint faying surfaces confirmed the decrease in slip resistance qualitatively. Although no direct relation is known between surface roughness and the slip coefficient, it is well known that a smooth surface results in a low slip coefficient. The tests reported in Ref. 9.2 showed a significant decrease in surface roughness of the two component zinc dust paint treated specimens compared to other types of surface treatments. It is also apparent that a wide range in the slip coefficient will result for surfaces treated with organic zinc-rich paints.

Studies on coated surfaces have indicated that inorganic zinc-rich paints provide better slip resistance than zinc paints that use organic binding agents.[9.2, 12.1, 12.6] When zinc silicate paint has been used with a clear lacquer (water-glass) as a binding agent and zinc dust powder as the pigment, high slip resistance has resulted. The increased hardness of the zinc silicate coat-

Table 12.6. Slip Coefficients for Surfaces Treated with Zinc Silicate Paint[a]

Coating Thickness (mils)	Average Slip Coefficient Product Name				
	A	B	C	D	E
0.8	0.41	0.47	0.62	0.53	0.50
2.0	0.52	0.53	0.64	0.56	0.52

[a] Specimens were blast cleaned and treated with different zinc silicate paints. These results are averaged from two readings each (see Ref. 12.6).

Table 12.7. Comprison of Slip Coefficients of Blast-Cleaned Surfaces and Coated Surfaces (Zinc Silicate Paint)

Ref.	Type of Treatment	Coating Thickness (mils)	Number of Tests	Average	Standard Deviation
12.1	1. Grit blasted	0.8–1.6	10	0.572	.071
	2. Zinc silicate paint	1.6–3.8	10	0.684	.047
	Shot blasted	—		0.600	—
	1. Shot blasted	3.0	—	0.630	—
	2. Zinc silicate paint				
	Grit blasted	—		0.580	—
	1. Grit blasted	3.0	—	0.560	—
	2. Zinc silicate paint				
9.2	Sand blasted	—	4	0.603	—
	1. Sand blasted	1.4	6	0.607	.006
	2. Zinc silicate paint	1.2	6	0.588	.021
		1.2	6	0.524	.012
		1.0	6	0.534	.013
		1.0	6	0.577	.035
		1.0	6	0.600	.008

ing provides a more slip-resistant surface than surfaces treated with organic zinc-rich paints. For optimum results these paints are generally applied to blast-cleaned surfaces by either spraying or brushing.[12.12]

The thickness of zinc silicate coatings also slightly influences the slip coefficient. This is illustrated in Table 12.6 where test results for different coating thickness are summarized.[12.6] The specimens were all blast cleaned and then coated with zinc silicate paint, supplied by five different suppliers. An increase in coating thickness increased the slip resistance for all five mixes.

Tests were also performed in Germany on sand-blasted specimens treated with zinc silicate paint.[9.2] The results are compared with plain sand-blasted surfaces in Table 12.7. The zinc paint was provided by five different suppliers and the coating thickness varied from 0.001 to 0.0015 in. All five coatings provided slip coefficients which were about the same as plain sand-blasted surfaces. The maximum difference in average slip coefficients between coated and uncoated specimens was about 12%. The results also indicate that the chemical composition of the paint does not greatly influence the slip behavior. Much greater variation was observed with organic

12.3 JOINT BEHAVIOR UNDER SUSTAINED LOADING

zinc-rich paints. Blast cleaned surfaces treated with zinc silicate paints are likely to yield a slip coefficient which is about the same as the slip coefficient provided by blast-cleaned base metal.

12.3 JOINT BEHAVIOR UNDER SUSTAINED LOADING

Field experience and test results have indicated that galvanized members may have a tendency to continue to slip under sustained loading.[12.3, 12.6, 12.9] Slip is stopped when the bolts come into bearing. In some situations this small slippage may impair the serviceability of the structure. Hence if a joint is subjected to sustained loading conditions and is slip critical, the slip performance of the coating must be considered under the sustained load condition.

Laboratory tests have been performed to evaluate the load-deformation behavior of different types of coated surfaces subjected to sustained loading.[12.3, 12.6, 12.9] In general, the observed slip behavior with respect to time can be characterized by one of the three relationships shown in Fig. 12.2.[12.9] Curve 1 represents a class of connections in which major slip occurs during application of the load. The bolts come into bearing against the

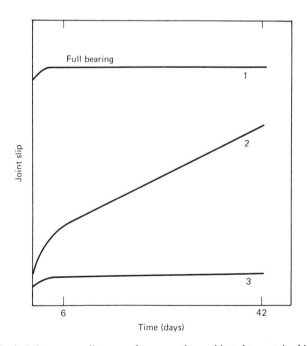

Fig. 12.2. Typical time versus slip curves for connections subjected to sustained loading.

plate and the joint remains stable with time unless the load reverses direction. Curve 2 represents connections which do not initially slip into bearing but continue to slip under sustained loading (the connection creeps). The slip rate under sustained loading only becomes zero when the bolts come into bearing. Curve 3 shows good slip resistance under short term as well as sustained loading conditions. After a small initial extension, often elastic, no further slip is detected.

Test specimens are usually subjected to stepwise increasing loads when evaluating slip resistance under sustained loading. After slip has been arrested or the slip rate has stabilized, the load is increased. This process is repeated until either the bolts are in bearing or the slip resistance of the faying surfaces is exceeded.

Tests on hot-dip galvanized joints subjected to sustained loading show a steady-state rate of slip.[12.6, 12.9] The connections developed a creep-type behavior as indicated in Fig. 12.2. Preassembly treatments which yielded an increase in short-duration slip resistance did not significantly improve the slip behavior under sustained loading. Joints treated with organic zinc-rich paints showed essentially the same behavior as galvanized joints.[9.1] The zinc layer created by the organic zinc-rich paint acts like a lubricant between the surfaces and this results in creep under sustained loading.

Better results were obtained using zinc silicate paint on the joint faying surfaces. Both short-duration slip resistance and sustained load slip resistance were improved. Test results indicated that a coating layer thickness equal to 0.002 to 0.0025 in. (50 to 60 μm) provided about the same slip coefficient for sustained loading and short duration tests.[12.3, 12.6] Even when the sustained loads were close to the slip load of the connection, a stable joint condition resulted.

Metallizing with either zinc or aluminum resulted in good short-duration slip resistance. However, under sustained loading conditions, aluminum sprayed faying surfaces provided better slip resistance than zinc-sprayed surfaces. Slip coefficients for sprayed aluminum surfaces were found to be about the same for both the sustained and short-duration loading tests. Zinc-sprayed surfaces exhibited creep when the joint was subjected to loads which were close to the slip resistance of the surfaces. If an appropriate factor of safety was applied so that the loads were well below the slip resistance of the joint, satisfactory sustained load characteristics were observed.[12.3, 12.6]

12.4 JOINT BEHAVIOR UNDER REPEATED LOADING

The behavior of plain, noncoated bolted butt joints, subjected to repeated loading conditions, is summarized in Chapter 5. For slip-resistant joints,

12.4 Joint Behavior Under Repeated Loading

crack initiation and growth were generally observed to occur through the gross section. When the slip resistance was decreased, failure usually occurred at the net section. The application of a protective surface coating may alter the slip resistance of a joint; therefore, its influence on the fatigue strength of a joint has to be examined.

Fatigue tests have been performed on hot-dip galvanized joints because they exhibited low slip resistance during short-duration static slip tests.[4.11, 4.27, 12.10, 12.13] Other protective surface treatments such as metallizing and zinc-rich paints have been studied as well.[5.17, 12.4, 12.6] The results of these tests have indicated that the lower slip resistance and early slips in the joints did not influence the fatigue resistance of coated joints. Their fatigue strength was equal or greater than the fatigue resistance of uncoated joints of similar dimensions.

In an attempt to explain qualitatively this behavior, joints were classified into two categories depending on whether the slip resistance of the test joint was exceeded by the applied load. It is shown in Chapter 5 that uncoated slip-resistant bolted joints subjected to repeated loading exhibit a fretting-type crack initiation in the gross section ahead of the bolt hole. Hence a surface coating which provides sufficient slip resistance should provide comparable behavior. Such behavior was frequently observed in tests on metallized and zinc silicate painted joints.[9.2, 12.4] For these surface conditions, repeated cyclic loads close to the slip load of the connection did not result in significant slip in the connection.

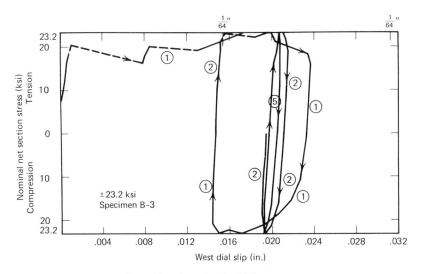

Fig. 12.3. "Lock-up" effect of hot-dip galvanized joints.

Tests of hot-dip glavanized joints showed that the connection either slipped into bearing or the connection "locked-up" (ceased to slip) after a few cycles when there was load reversal.[4.11, 12.10, 12.13] This locking-up effect is illustrated in Fig. 12.3 for a hot-dip galvanized joint subjected to repeated load reversal. Figure 12.3 shows that the displacements during the fifth cycle were about the same as the second cycle. Hence small slips in hot-dip galvanized joints did not decrease the fatigue life. Failures often occurred through the gross section area despite the initial slip.

Disassembly of joints confirmed the tendency to lock up. To separate the plates of a joint it was often necessary to pry them apart.[4.11] Layers of zinc tended to pull off from the surfaces of the plate as a result of galling and seizing of the zinc coating in the region around the bolt holes where high contact pressures exist.

A preassembly treatment of the hot-dip galvanized surface by wire brushing or light blast cleaning did not influence the fatigue life.

12.5 DESIGN RECOMMENDATIONS

Joints with metallic coatings should be designed by the criteria suggested in Chapter 5. Depending on the type of surface treatment, a wide range of slip coefficients are possible. Even for a specific type of treatment substantial scatter can result from fabrication procedures.

Allowable shear stresses can be approximated from the slip coefficients available for the surface treatments summarized in this chapter. A lower bound slip coefficient for different surface conditions was selected and β_1 values and allowable shear stresses determined. Table 12.8 summarizes the results. It also shows the "average" lower bound slip coefficient and the standard deviation for different surface treatments. Available test data on specific surface treatments may permit an increase in these design values, since the suggested values are conservative estimates.

When slip-resistant joints are subjected to sustained loading conditions, only surface treatments which provide adequate slip resistance under long-term loading should be used. Metallizing with either zinc or aluminum or a zinc silicate paint should be used. Hot-dip galvanizing and organic zinc-rich paint systems are not satisfactory for slip-resistant joints. Obviously ribbed bearing bolts would be satisfactory for these conditions as they would not permit substantial slips to develop.

If a joint is subjected to repeated loads, the design recommendations given in Chapter 5.4 are applicable. If the slip resistance is adequate to prevent slip during the lifetime of the structure, the stress range on the gross section area may be used for design. If slip is expected, the design stress range should be applied to the net section. Although several hot-dip

Table 12.8. Reduction Factors β_1 for Evaluation of Design Shear Stresses for Slip-Resistant Coated Surfaces.[a]

	Slip Coefficient		Slip Probability			
			A325		A490	
Surface Treatment	Average	Standard Deviation	5%	10%	5%	10%
Hot-dip galvanized	0.18	.040	0.31 (9.3)	0.36 (10.8)	0.27 (10.9)	0.31 (12.6)
Hot-dip galvanized, treated, wire brushed or blasted	0.40	.070	0.76 (22.9)	0.85 (25.7)	0.66 (26.7)	0.76 (30.4)
Vinyl treated	0.27	.023	0.62 (18.7)	0.66 (19.9)	0.54 (21.8)	0.57 (23.1)
Blast-cleaned zinc-sprayed ($t \geq$ 0.002 in.)	0.40	.040	0.88 (26.5)	0.94 (28.3)	0.77 (31.0)	0.82 (33.0)
Blast-cleaned Al-sprayed ($t \geq$ 0.002 in.)	0.55	.055	1.21 (36.5)	1.30 (39.0)	1.06 (42.6)	1.13 (45.3)
Blast-cleaned organic zinc-rich paint	0.35	.035	0.77 (23.2)	0.83 (24.8)	0.67 (27.1)	0.72 (28.8)
Blast-cleaned zinc-silicate paint	0.50	.050	1.10 (33.2)	1.17 (35.3)	0.96 (38.7)	1.03 (41.2)

[a] Summary of β_1 values. The number in parenthesis represents the allowable shear stress for $\beta_2 = 1.0$ (turn-of-nut method) and $\beta_3 = 1.0$ (see Chapter 5).

galvanized joints have exhibited gross section failures, it is recommended that these connections be designed on the basis of their net section area.

Since the presence of a coating does not affect the strength of a joint, the design recommendations given in Chapter 5 for joints that are not slip critical can be applied to all types of coated joints as well.

References

12.1 Office of Research and Experiments of the International Union of Railways (ORE), *Coefficients of Friction of Faying Surfaces Subjected to Various Corrosion Protective Treatments,* Report 2, ORE, Utrecht, The Netherlands, June 1967.

12.2 ORE, *Effects of Weathering on the Coefficients of Friction of Unprotected and Protected Faying Surfaces,* Report 3, ORE, Utrecht, The Netherlands, October 1968.

12.3 ORE, *Influence of Sustained Loading on the Slip Behavior of High Strength Bolted Joints,* Report 4, ORE, Utrecht, The Netherlands, October 1969.

12.4 ORE, *Influence of Coated Surfaces on the Fatigue Strength of High Strength Bolted Joints,* Report 5, ORE, Utrecht, The Netherlands, October 1970.

12.5 Centre de Recherches Scientifiques et Techniques de L'Industrie des Fabrications Metallique (CRIF), Section Construction Metallique, *Les Assemblages par Boulons de Haute Resistance,* Report MT 48, MT50, Brussels, Belgium, 1969 (in French).

12.6 T. v. d. Schaaf, *Influence of Protective Surface Coatings on the Slip Behavior of High Strength Bolted Joints Subjected to Sustained Loading,* Stevin Laboratory, Report 6-68-5-VB-18, Delft University of Technology, Delft, The Netherlands, 1968 (in Dutch).

12.7 S. Y. Beano and D. D. Vasarhelyi, *The Effect of Various Treatments of the Faying Surface on the Coefficient of Friction in Bolted Joints,* University of Washington, Department of Civil Engineering Seattle, December 1958.

12.8 L. Zennaro, *Slip Tests of High Strength Bolted Joints with Different Galvanized Coating Structures,* Document CECM-X-71-8.

12.9 V. Lobb and F. Stoller, "Bolted Joints Under Sustained Loading," *Journal of the Structural Division, ASCE,* Vol. 97, ST3, March 1971.

12.10 D. J. L. Kennedy, *High Strength Bolted Galvanized Joints,* Engineering Extension Series 15, Proceedings, ASCE Specialty Conference on Steel Structures, University of Missouri, Columbia, 1970.

12.11 American Welding Society, *Recommended Practices for Metallizing with Aluminum and Zinc for Protection of Iron and Steel,* American Welding Society C2.2-67, New York, 1967.

12.12 Steel Structures Painting Council (SSPC), *Guide to Zinc-Rich Coating Systems,* SSPC Specification 12.00.

12.13 W. H. Munse and P. C. Birkemoe, *High Strength Bolting of Galvanized Connections,* The Australian Institute of Steel Construction and the Australian Zinc Development Association, Sydney-Melbourne, Australia, August 1969.

Chapter Thirteen

Eccentrically Loaded Joints

13.1 INTRODUCTION

In eccentrically loaded joints the connection is subjected to applied loads that result in a line of action passing outside the center of rotation of the fastener group. Some common examples are bracket-type connections, web splices in beams and girders, and the standard beam connections shown in Fig. 13.1. Due to the eccentricity of the applied load, the fastener group is subjected to a shear force and a twisting moment. Both the moment and the shear force result in shear stresses in the fasteners. Hence the governing shear stress in each fastener is the resultant of two components, one caused by the applied shear force and the other resulting from the moment due to the eccentricity of the load.

Usually the centric shear force is assumed to be equally distributed among the fasteners. The evaluation of the influence of the moment on the bolt shear stresses is more complex and was studied elastically as early as 1870.[13.1] Recent research has yielded information on the ultimate strength of specific types of connection.[13.2, 13.3, 13.4, 13.6]

This chapter deals with the analysis and design of eccentrically loaded fastener groups. Emphasis will be placed upon the design of a connection as shown in Fig. 13.1a. The application to web splices in girders (Fig. 13.1b) and standard beam connections is discussed in Chapters 16 and 18, respectively.

13.2 BEHAVIOR OF A FASTENER GROUP UNDER ECCENTRIC LOADING

Tests on special connections have been performed to evaluate the load-deformation behavior of fastener groups subjected to an eccentric shear load. Riveted as well as high-strength bolted connections have been examined.[13.2-13.5] All test specimens were of the type shown in Fig. 13.2 with a fastener group consisting of one or two vertical lines of fasteners. Since the connection is symmetric with respect to the line of action of the load, each test provides two load deformation curves for identical connections.

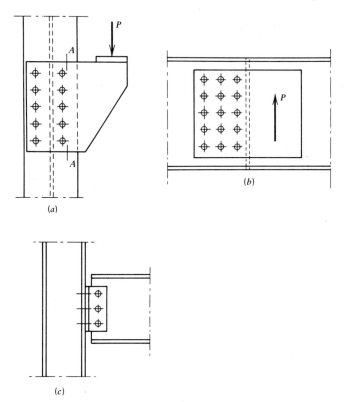

Fig. 13.1. Typical eccentrically loaded connections. (a) Bracket connection; (b) Beam-girder web splice; (c) standard beam connection.

In general, the design of the test specimens caused the fasteners in the web angles to be the critical components. Since the fastener group was the critical component, the test results can be used to assess the strength of the fastener group. However, the load-deformation behavior of a connection in the field may also be affected by other components of the connection as well.

The behavior of various fastener patterns under different eccentricities can be represented by load rotation curves as shown in Fig. 13.3.[13.4] The straight line from the origin to point A represents the elastic rotation. The transition segment AB represents elastic as well as plastic deformations. Beyond point B the rotation is mainly produced by plastic deformations. This segment of the load-rotation curve is terminated by the failure load which is reached as one or more of the fasteners fail in shear.

13.2 Behavior of a Fastener Group under Eccentric Loading

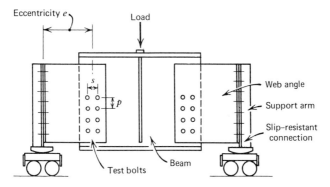

Fig. 13.2. Test specimen with eccentrically loaded fastener group.

Load-rotation curves have been developed for riveted as well as bolted specimens.[13.2-13.4] Figure 13.4 shows a typical load rotation curve from Ref. 13.2 for a bolted specimen with two vertical rows of $\frac{3}{4}$-in. A325 bolts. The horizontal distance from the load to the centroid of the fastener group was equal to 12 in. In this test series the bolt holes in the beam web and web angles were match drilled for fitted bolts. The resulting minimum clearance between the bolt and the hole minimized the joint slip. In practice, bolts are usually placed in holes with $\frac{1}{16}$ in. clearance. If hole clearance is present, slip may occur when the slip resistance of the connection is exceeded. Slip will bring one or more fasteners into bearing. Thereafter, the connection will behave in much the same way as described by Fig. 13.4.

The amount of slip to be expected depends on the hole clearance, the fastener pattern, and the alignment of the holes in the connection. The rotation due to slip decreases rapidly with an increase in distance from the outermost fastener to the center of rotation of the bolt group. In most practical situations the slips will be so small that they do not have a signifi-

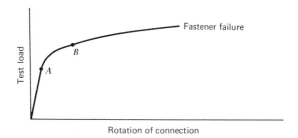

Fig. 13.3. Idealized load-rotation diagram for an eccentrically loaded fastener group.

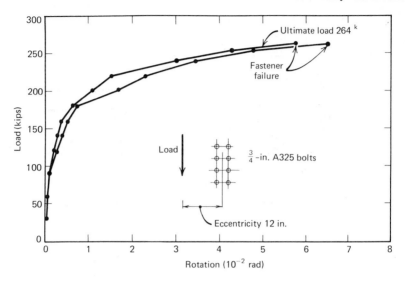

Fig. 13.4. Load-rotation curve for bolted connection (Ref. 13.2). (Test specimen shown in Fig. 13.2.)

cant effect on the serviceability of the structure. Therefore, most joints can be designed on the basis of the ultimate strength of the joint.

13.3 ANALYSIS OF ECCENTRICALLY LOADED FASTENER GROUPS

For many years the analysis and design of eccentrically loaded fastener groups was based on the assumption that the fasteners behaved elastically and were not stressed beyond the proportional limit.[13.10, 13.11] The eccentric load was resolved into a shear load P acting through the centroid of the fastener group and a moment Pe, where e is the eccentricity of P with respect to the centroid of the fastener group. The shear force acting through the centroid was assumed to be distributed uniformly among the fasteners as in other shear splices. The moment was assumed to cause stresses in the fasteners which vary linearly with the distance from the fastener to the center of rotation. Because elastic behavior of the fasteners was assumed, the center of rotation for evaluating the shear component of a fastener due to the moment coincided with the centroid of the fastener group. The resulting stress in the fastener was evaluated by vectorially adding the stress from each load component, that is, the centric shear force and the moment. The method further assumed the connected plates to be rigid enough to remain essentially undeformed during twist and results in a

13.3 Analysis of Eccentrically Loaded Fastener Groups

linear strain variation for the fasteners. The influence of the frictional resistance between the component parts and the load-deformation capacity of the fasteners is neglected.

Tests on eccentrically loaded riveted connections indicated that the elastic analysis yielded a conservative design.[13.4, 13.9] On the basis of test results the method was modified by introducing an "effective eccentricity," which is less than the actual eccentricity. Empirical formulas to determine the effective eccentricity as a function of specific fastener patterns were developed.[13.4, 13.7] Reduction in eccentricity yielded a factor of safety more compatible to the value used for shear alone. The method is essentially based on the elastic behavior of the fastener group described in this section. Reducing the eccentricity decreases the magnitude of the bending component and recognizes the actual strength of the joint observed in tests.[13.4, 13.9]

Although the use of either the effective or full eccentricity has provided safe designs, the factor of safety with respect to ultimate load is still variable although the use of the effective eccentricity reduces this variability. Neither method takes full advantage of the deformation capacity of the fastener. Recently a rational method for predicting the ultimate strength of an eccentrically loaded fastener group has been developed which considers the complete load-deformation relationship of a single fastener.[13.2] With a slight modification this method of analysis is also applicable to slip-resistant joints.

13.3.1 Slip-Resistant Joints

Initially the load-deformation curve of an eccentrically loaded joint is approximated by a straight line, representing the elastic rotation. During this stage the applied load is completely carried by frictional resistance between the constituent parts of the connection. This phase of load transfer is generally terminated by slip of the connection. The load at which the slip-resistance of the fastener group is exceeded causes movement and brings one or more bolts into bearing.

The slip load can be approximated by considering the following assumptions:

1. At the slip load, the connection rotates about an instantaneous center of rotation.
2. At the slip load of the connection, the maximum slip resistance of each individual fastener is reached. An analogous assumption has been used to describe the slip resistance of simple shear splices.
3. The slip resistance of the individual fasteners can be represented by a force at the center of the bolt acting perpendicular to the radius of rotation.

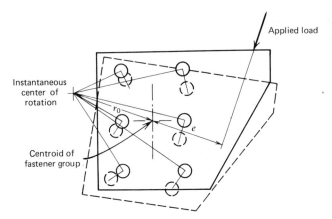

Fig. 13.5. Instantaneous center of rotation.

The instantaneous center is defined as follows. The eccentric load on the connection produces a rotation about the centroid of the fastener group together with a translation of one plate with respect to the other. The rotation and translation can be reduced to a pure rotation about a point defined as the instantaneous center of rotation (see Fig. 13.5). The location of this point depends on the fastener arrangement as well as on the direction and point of application of the applied load.

The maximum slip resistance R_S of a single fastener was described in Chapter 5 and can be expressed as

$$R_S = mk_s T_i \qquad (13.1)$$

Therefore, based on the previously stated assumptions, at the slip load each fastener is subjected to a load R_S acting perpendicular to the radius of rotation. Figure 13.6 shows schematically the load transfer for a symmetric fastener pattern. The three equations of equilibrium can be employed to determine the coordinates of the instantaneous center and the maximum value of the load which results in slip of the connection. The solution of this problem is generally accomplished by an iterative procedure. A trial location of the instantaneous center can be selected. For convenience, the origin of the coordinate system can be placed at the instantaneous center with the x-axis perpendicular to the applied load. The radius of rotation r_i of the ith fastener is equal to

$$r_i = \sqrt{x_i^2 + y_i^2} \qquad (13.2)$$

13.3 Analysis of Eccentrically Loaded Fastener Groups

Equating the sum of all forces in the x and y direction as well as the sum of the moments about the instantaneous center to zero, yields

$$\sum_{i=1}^{n} R_S \sin \varphi_i = 0 \tag{13.3}$$

$$\sum_{i=1}^{n} R_S \cos \varphi_i - P = 0 \tag{13.4}$$

$$P(e + r_0) - \sum_{i=1}^{n} r_i R_S = 0 \tag{13.5}$$

Equations 13.3 and 13.4 are usually written as follows

$$R_S \sum_{i=1}^{n} \frac{y_i}{r_i} = 0 \tag{13.6}$$

$$R_S \sum_{i=1}^{n} \frac{x_i}{r_i} - P = 0 \tag{13.7}$$

The solution to the problem is achieved if the value of r_0 satisfies all three equilibrium equations. The procedure can be repeated until this condition is met.

A symmetric fastener pattern was used in Fig. 13.6 and the applied load was normal to the axis of symmetry. In such situations the instantaneous center of rotation must lie on the axis perpendicular to the applied load to satisfy Eq. 13.7. The procedure also applies to the more general case where

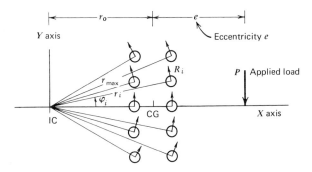

Fig. 13.6. Analyses of eccentrically loaded fastener group. CG: Center of gravity of fastener group. IC: Instantaneous center of rotation. For slip-resistant joints R_i is equal to R_s where $R_s = mK_sT_i$). For other joints $R_i = (r_i/r_{max}) R_{ult}$.

no axis of symmetry of the fastener group exists or the applied load is acting in an arbitrary direction, as in Fig. 13.5.

It is also apparent from Eqs. 13.5, 13.6, and 13.7 that if the instantaneous center and the centroid of the fastener group coincide at the location of a fastener, this fastener in general cannot be fully effective as far as slip resistance is concerned. Hence care must be exercised in assuming that all fasteners have maximum resistance when large eccentricities exist.

13.3.2 Ultimate Strength Analysis

A theoretical approach to predict the ultimate strength of an eccentrically loaded fastener group was developed by Crawford and Kulak.[13.2] This approach considers the load-deformation response of a single fastener as a basis for determining the ultimate strength of a fastener group. The method proposed by Crawford and Kulak utilizes the load deformation behavior of a single fastener loaded in double shear. This relationship has been expressed as [5.22]

$$R = R_{ult}(1 - e^{-\mu\Delta})^\lambda \tag{13.8}$$

in which R = shear force on the bolt
R_{ult} = the ultimate shear load of the fastener
Δ = the shearing, bending, and bearing deformation of the fastener as well as the local bearing deformation of the connecting plates
μ, λ = regression coefficients
e = base of natural logarithms

Numerical values for R_{ult}, λ and μ for various combinations of bolts and connected material can be determined experimentally by means of special shear tests. A tension-type shear test has been recommended, since it yields a lower bound to the ultimate shear capacity R_{ult} of the bolt.[4.4]

The evaluation of the ultimate strength of an eccentrically loaded fastener group is comparable to the analysis of similar slip-resistant joints. The connection is assumed to rotate about an instantaneous center and the connected plates are assumed to remain rigid during this rotation. The latter assumption implies that the deformation occurring at each fastener varies linearly with its distance from the instantaneous center. The fastener deformation and the resulting shear load on the fastener is acting perpendicular to the radius of rotation of the fastener. The ultimate strength of the fastener group is assumed to be reached when the ultimate strength of the fastener farthest away from the instantaneous center is reached.

For a given fastener configuration and an eccentricity of the load equal to e, a trial location of the instantaneous center can be selected at a dis-

13.3 Analysis of Eccentrically Loaded Fastener Groups

tance r_0 from the centroid of the fastener group (see Fig. 13.6). The radius of rotation r_i of the ith fastener is given by Eq. 13.2. At ultimate load, the shear deformation of the critical fastener, located a distance r_{max} away from the instantaneous center, is assumed equal to Δ_{max}, the maximum fastener deformation obtained from a single bolt shear test.[4.4, 5.22] The deformation of other fasteners can be determined from

$$\Delta_i = \frac{r_i}{r_{max}} \Delta_{max} \tag{13.9}$$

The fastener load corresponding to Δ_i is readily obtained from Eq. 13.8.

Equilibrium of horizontal and vertical forces yields

$$\sum F_x = 0; \quad \sum_{i=1}^{n} R_i \sin \varphi_i = 0 \tag{13.10}$$

$$\sum F_y = 0; \quad \sum_{i=1}^{n} R_i \cos \varphi_i - P = 0 \tag{13.11}$$

The summation of moments around the instantaneous center yields a third equation

$$P(e + r_0) - \sum_{i=1}^{n} r_i R_i = 0 \tag{13.12}$$

Equations 13.10 and 13.11 can be conveniently written in terms of the coordinates x_i, y_i of the fastener,

$$\sum_{i=1}^{n} \frac{R_i y_i}{r_i} = 0 \tag{13.13}$$

$$\sum_{i=1}^{n} \frac{R_i x_i}{r_i} - P = 0 \tag{13.14}$$

The solution is obtained when the estimated value of r_0 satisfies Eqs. 13.12, 13.13, and 13.14 simultaneously.

13.4 COMPARISON OF ANALYTICAL AND EXPERIMENTAL RESULTS

The validity of the ultimate strength analysis has been checked by comparing predicted results with experimental data. It was found that the predicted ultimate loads for bolted specimens ranged between 5 and 14% higher than the observed failure loads of the connections.[13.2]

One of the reasons for this observed difference is because the deformation of the critical fastener in the connection does not reach the maximum value observed in a single bolt shear test. In the single bolt test the load and deformation direction do not change. In the eccentrically loaded connection, the load and deformation of each bolt is changing direction continuously as the instantaneous center moves with an increase in applied load. It was observed from tested specimens that the bolt holes were deformed and scored by the circular movement of the bolts. Hence it is unlikely that the critical fastener in the connection will deform as much as a single fastener loaded with a unidirectional force.[13.2]

The predictions of the ultimate strength in Ref. 13.2 were based on load-deformation relationships determined from compression type specimens. However, failure of the fasteners was observed mainly in the tension region of the plates (where the connected plates are subjected to tension). It was shown in Chapter 4 that a tension-type shear test generally yields lower shear values than a compression-type shear test.[4.4] Since a compression type shear test was used by Crawford and Kulak, this may also have contributed to the overestimation of the ultimate loads of the bolt groups reported in Ref. 13.2.

At the present time (1973), little information on the slip-behavior of eccentrically loaded fastener groups is available.[13.12] Additional test data are required to make a feasible comparison between the experimental and analytical data.

Although the ultimate strength of an eccentrically loaded fastener group of a type as given in Figs. 13.1a or 13.2 can be evaluated within acceptable limits, additional research is needed to be able to predict the load deformation behavior of such joints. Furthermore, research on other types of connections, such as shown in Fig. 13.1b is desirable to verify the application of the analysis as outlined in the previous section to these types of connections as well.

13.5 DESIGN RECOMMENDATIONS

13.5.1 Connected Material

The design of the plates used in eccentrically loaded joints does not involve special design recommendations. To design the plate for the bracket connection shown in Fig. 13.1a, the shear stress and normal stress at section AA due to the applied load P should be checked. If relatively thin plates are used, the out-of-plane deformations due to instability effects may require an increased plate thickness.

The allowable stresses for these conditions depend on the plate material and the type of loading.

13.5 Design Recommendations

13.5.2 Fasteners

The fasteners in an eccentrically loaded connection can be designed on the basis of the load capacity predicted from the instantaneous center method. Depending on the required joint performance, the design load of a joint is based on either the slip load or the ultimate strength of the connection. Both require the determination of the instantaneous center of rotation of the fastener group. Solutions for typical fastener configurations subjected to varying amounts of eccentricity have been obtained. These results can be presented as load tables or by means of design charts. For practical reasons both load tables and design charts are usually limited to the most commonly used fastener patterns. Unusual fastener patterns can be evaluated however.

i. **Allowable Stress Design Bolted Joints.** The ultimate strength of an eccentrically loaded connection can be used to develop design loads. The ultimate strength of the connection was defined as the load at which the shear strength τ_u of one or more fasteners was reached. By specifying an allowable shear stress for the fastener, an allowable load can be developed. The ultimate load P_u of a connection can be expressed in terms of the shear area A_{SH} of a typical fastener and a factor k which depends on factors such as the bolt grade, joint dimensions, fastener pattern, and load eccentricity.

$$P_u = kA_{SH} \tag{13.15}$$

To separate the effects of bolt grade and geometrical factors it is convenient to write k as follows

$$k = C\tau_u \tag{13.16}$$

where the shear strength τ_u of the bolt reflects the influence of the bolt grade and the factor C depends on geometrical conditions only. Combining Eqs. 13.15 and 13.16 gives

$$P_u = CA_{SH}\tau_u \tag{13.17}$$

The product $A_{SH}\tau_u$ represents the shear capacity of a typical fastener. Once the factor C is known for a particular fastener pattern, the allowable load P_a can be determined by applying an appropriate factor of safety F with respect to the ultimate strength of the joint. This results in

$$P_a = \frac{CA_{SH}\tau_u}{F} \tag{13.18}$$

The factor τ_u/F is directly analogous to the allowable shear stress τ_a developed for simple shear splices. Hence

$$P_a = CA_{SH}\tau_a \tag{13.19}$$

The factor C must be determined from an ultimate strength analysis of the joint in question.

Kulak and Crawford analyzed a great number of hypothetical joints and developed ultimate load tables for connections with one or two lines of fasteners with up to 12 bolts per line.[13.2] The load P was applied at varying eccentricities. In order that fastener configurations and/or eccentricities which are not tabulated can be considered, polynomial functions were fitted to the results. The polynomials provided a good approximation of the ultimate load and permitted extrapolation to other geometric conditions. For one and two lines of fasteners the factor k was approximated by the relationship

$$k = \alpha I^\beta \tag{13.20}$$

where I is the relative moment of inertia of the bolt group for unit bolt areas. (For one fastener line $I = I_x$; for two or more fastener lines $I = I_x + I_y$.) The parameters α and β are polynomial functions of the eccentricity e.

For one line of A490 bolts, with the load eccentricity e taken in inches, the coefficients α and β were evaluated as

$$\alpha = 1.02 + \frac{61.5}{e} + \frac{464}{e^2} - \frac{664}{e^3} \tag{13.21}$$

$$\beta = 0.645 - \frac{0.129}{e} - \frac{3.85}{e^2} + \frac{7.43}{e^3} \tag{13.22}$$

For two or more lines of A490 fasteners the coefficients in Eq. 13.21 become 1.23, 80.1, 546, and 808. The coefficients for β become 0.651, 0.183, 3.13, and 6.25, respectively. The eccentricity e is the distance between the applied load and the centroid of the fastener group. A comparison between the theoretical ultimate loads of a connection and the ultimate loads as predicted by the fitted approximation indicated that the ultimate loads were estimated to within 5% of the predicted strength. In a few extreme cases the error increased to 10%.

The mathematical approximations were determined from the predicted strengths of hypothetical joints fastened by minimum strength A490 bolts. The C value for Eqs. 13.17, 13.18, and 13.19 can be evaluated from Eqs. 13.16 and 13.20. C values for joints with one or two lines of fasteners and a varying eccentricity of the load are listed in Tables 13.1 and 13.2. In these tables the parameter α' is defined as α/τ_u. The convenience of a conversion becomes apparent by examining Eqs. 13.16 and 13.20. The shear strength τ_u was taken as $0.65\sigma_u$. This provided a mean value between the

13.5 Design Recommendations

Table 13.1. C-Values for One Line of Fasteners[a]

e	\multicolumn{11}{c}{Number of Fasteners}
	2	3	4	5	6	7	8	9	10	11	12
3.00	0.97	1.81	2.73	3.72	4.79	5.91	7.09	8.33	9.61	10.94	12.30
4.00	0.74	1.46	2.28	3.20	4.21	5.29	6.45	7.68	8.97	10.32	11.73
5.00	0.59	1.23	1.99	2.86	3.84	4.91	6.08	7.33	8.66	10.07	11.55
6.00	0.49	1.05	1.74	2.56	3.48	4.51	5.64	6.86	8.18	9.58	11.07
7.00	0.41	0.91	1.54	2.29	3.15	4.11	5.18	6.35	7.61	8.96	10.41
8.00	0.36	0.80	1.37	2.06	2.85	3.75	4.75	5.84	7.04	8.32	9.70
9.00	0.31	0.72	1.23	1.86	2.59	3.42	4.36	5.38	6.50	7.71	9.01
10.00	0.28	0.64	1.12	1.69	2.37	3.14	4.01	4.97	6.02	7.15	8.38
11.00	0.25	0.58	1.02	1.55	2.18	2.90	3.71	4.61	5.59	6.66	7.80
12.00	0.23	0.54	0.94	1.43	2.02	2.69	3.45	4.29	5.21	6.21	7.29
14.00	0.19	0.46	0.81	1.24	1.75	2.35	3.02	3.76	4.58	5.47	6.44
16.00	0.17	0.40	0.71	1.10	1.55	2.08	2.68	3.35	4.09	4.89	5.76
18.00	0.15	0.36	0.64	0.98	1.40	1.88	2.42	3.03	3.70	4.43	5.22
20.00	0.14	0.32	0.58	0.89	1.27	1.71	2.21	2.77	3.38	4.05	4.78
22.00	0.12	0.30	0.53	0.82	1.17	1.58	2.04	2.55	3.12	3.74	4.42
24.00	0.11	0.28	0.49	0.76	1.09	1.46	1.89	2.37	2.90	3.48	4.11

[a] In general $C = \alpha' I^\beta$ where $I = I_x + I_y$

$$\alpha' = 0.0104 + \frac{0.625}{e} + \frac{4.719}{e^2} - \frac{6.750}{e^3}$$

$$\beta = 0.645 - \frac{0.129}{e} - \frac{3.85}{e^2} + \frac{7.43}{e^3}$$

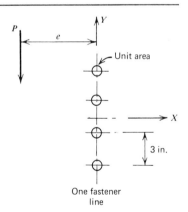

average shear strengths determined from both tension-type and compression-type shear tests. The mean value was used since both types of shear loading occur simultaneously in an eccentrically loaded joint.

With the tabulated C values the allowable load for a joint can be obtained by multiplying the tabulated C value by the allowable shear load for a fastener. The application to other grades of fasteners is conservative

Table 13.2. C Values for Two Lines of Fasteners[a]

	Number of Fasteners										
e	2	3	4	5	6	7	8	9	10	11	12
3.00	2.34	3.79	5.48	7.38	9.46	11.70	14.08	16.59	19.23	21.99	24.85
4.00	1.88	3.15	4.67	6.43	8.38	10.53	12.84	15.31	17.93	20.69	23.59
5.00	1.58	2.72	4.14	5.81	7.71	9.82	12.12	14.62	17.29	20.13	23.14
6.00	1.34	2.37	3.68	5.24	7.04	9.06	11.29	13.72	16.35	19.16	22.16
7.00	1.16	2.09	3.28	4.73	6.40	8.30	10.41	12.73	15.25	17.96	20.86
8.00	1.02	1.86	2.95	4.28	5.83	7.60	9.58	11.76	14.14	16.71	19.47
9.00	0.91	1.67	2.67	3.89	5.33	6.97	8.82	10.87	13.10	15.53	18.13
10.00	0.82	1.51	2.43	3.56	4.89	6.43	8.15	10.06	12.16	14.44	16.89
11.00	0.74	1.38	2.23	3.28	4.52	5.95	7.56	9.35	11.32	13.46	15.77
12.00	0.68	1.27	2.06	3.03	4.19	5.53	7.04	8.72	10.58	12.59	14.77
14.00	0.58	1.10	1.78	2.64	3.65	4.85	6.19	7.68	9.33	11.14	13.08
16.00	0.51	0.97	1.58	2.34	3.26	4.32	5.52	6.87	8.36	9.98	11.74
18.00	0.46	0.87	1.42	2.11	2.94	3.90	5.00	6.22	7.58	9.06	10.67
20.00	0.41	0.79	1.29	1.92	2.68	3.57	4.57	5.70	6.94	8.31	9.79
22.00	0.38	0.72	1.19	1.77	2.47	3.29	4.22	5.27	6.42	7.69	9.06
24.00	0.35	0.67	1.10	1.65	2.30	3.06	3.93	4.91	5.98	7.17	8.45

[a] In general $C = \alpha' I^\beta$, where $I = I_x + I_y$

$$\alpha' = 0.0125 + \frac{0.814}{e} + \frac{5.550}{e^2} - \frac{8.220}{e^3}$$

$$\beta = 0.651 - \frac{0.183}{e} - \frac{3.13}{e^2} + \frac{6.25}{e^3}$$

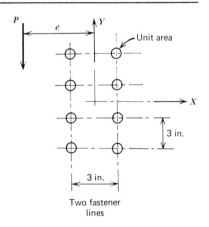

Two fastener lines

since the A490 bolt has the least ductility. Similar tables of C values can be developed for joints with other fastener patterns either by a theoretical analysis or by means of the mathematical expressions provided by Eqs. 13.20 through 13.22. For joints with more than two vertical lines of fasteners the expressions for α' and β as given in Table 13.2 might be acceptable in evaluating the fastener group coefficient C. Such a procedure is fol-

13.5 Design Recommendations

lowed in the method presently (1973) available in the *AISC Manual*.[13.7] A theoretical study to verify such an approximation is needed.

Load tables and additional mathematical expressions were developed for eccentrically loaded joints fastened by A325 bolts.[13.2] An examination of these data indicated that the capacity of A325 joints could be approximated with the fitted expressions for A490 bolts. A comparison between the allowable loads obtained from the ultimate strength analysis and the values obtained from Eq. 13.19 with $\tau_a = 30$ ksi in the critical fastener showed that the allowable loads were slightly conservative. The error introduced by applying the same C value for both A325 and A490 bolts for most joint configurations was small. For a few bolt patterns the capacity was underestimated by 10 to 12%.

The C values for rivets and A307 bolts can be developed on the basis of typical load-deformation curves. For convenience, the allowable loads for connections employing these fastener types can be conservatively estimated using the C value for high-strength bolts.

If the allowable shear stresses of 30 and 40 ksi for A325 and A490 bolts, respectively, are used to determine the working load capacity, the actual margin of safety for eccentrically loaded joints can be determined. Table 13.3 summarizes the experimental ultimate loads reported in Ref. 13.2.

Table 13.3. **Factor of Safety for Eccentrically Loaded Joints Fastened by A325 Bolts Designed by Different Methods**[a]

Spec.	P_u (Test) (kip)	Proposed Method		AISC Manual	
		P_{all}[b] (kip)	Factor Safety	P_{all}[c] (kip)	Factor Safety
B^1	225	89.0	2.52	84	2.68
B^2	230	90.2	2.55	75	3.09
B^3	190	72.1	2.63	60	3.17
B^4	251	99.5	2.52	79	3.18
B^5	221	88.0	2.52	67	3.28
B^6	264	106.8	2.48	82	3.10
B^7	212	86.4	2.46	63	3.36
B^8	266	106.0	2.52	76	3.42

[a] These tests are reported in Ref. 13.2; $\tfrac{3}{4}$-in. A325 bolts of minimum specified mechanical properties were used.

[b] According to design recommendations presented in Section 13.5, $\tau_{all} = 30$ ksi.

[c] According to design recommendation outlined in the *AISC Manual*, $\tau_{all} = 22$ ksi.

Table 13.4. Factor of Safety for Eccentrically Loaded Riveted Joints Designed by Different Methods[a]

Spec.	P_u (Test) (kip)	Proposed Method		AISC Manual	
		P_{all}[b] (kip)	Factor of Safety	P_{all}[c] (kips)	Factor Safety
TP-1	216	54.5	3.96	74	2.90
TP-2	161	43.5	3.79	60	2.69
TP-3	100	26.1	3.84	31	3.25
TP-4	550	138.2	3.96	159	3.46
TP-5	440	107.	4.10	150	2.93
TP-6	362	88.3	4.08	117	3.10
TP-7	222	52.2	4.24	58	3.83
TP-8	120	30.4	3.94	32	3.71
TP-9	568	131.5	4.30	184	3.09
TP-10	354	90.0	3.92	107	3.33

[a] These test results are reported in Refs. 13.3 and 13.4; A502 grade 1 rivets were used. Shear strength of the rivets was about 62 ksi.
[b] According to design recommendations presented in Section 13.5, $\tau_{\text{all}} = 15$ ksi.
[c] According to design recommendations presented in the *AISC Manual*, $\tau_{\text{all}} = 15$ ksi.

The results were determined from test specimens as shown in Fig. 13.2 with minimum strength A325 bolts. The allowable loads determined from Eq. 13.18 are also given in Table 13.3. The allowable design shear stress was taken as 30 ksi and C was taken from Tables 13.1 and 13.2. Based on these values the factor of safety is seen to vary from 2.43 to 2.63. For comparative purposes the allowable loads determined by the method outlined in the 7th edition of the *AISC Manual* is listed as well.[13.7] It is apparent that a more uniform factor of safety is provided by the proposed method.

Similar comparisons were made for riveted specimens (see Table 13.4). The shear strength of the rivets in the test specimens was equal to 62 ksi, which was considerably above the minimum strength of A502 Grade 1 rivets. If the allowable shear stress is taken as 15 ksi for the rivets the factor of safety is seen to vary from 3.84 to 4.30. Since the shear strength of rivets is more likely to be substantially lower than 62 ksi, the provided margin of safety is reasonable. Other tests on A502 Grade 1 rivets have indicated that the shear strength is about equal to 0.6–0.7 σ_u. Since the rivet material is more likely to have a tensile strength of about 60 ksi (see Chapter 3), the expected shear strength is about 40 ksi. It is believed this will provide factors of safety more compatible with bolted joints.

DESIGN RECOMMENDATIONS ECCENTRICALLY LOADED JOINTS

Allowable Stress Design

Shear Stresses for High-Strength Bolts

$$\tau_a = \tau_{basic}$$

where $\tau_{basic} = 30$ ksi for A325 bolts
$\tau_{basic} = 40$ ksi ksi for A490 bolts

Allowable Joint Loads
(a) Shear planes pass through bolt shank

$$P_a = CmA_b\tau_a$$

where m = number of shear planes
A_b = nominal bolt area
C = fastener group coefficient (see Tables 13.1 and 13.2)

(b) Shear planes pass through threads

$$P_a = 0.75CmA_b\tau_a$$

Bearing Stresses
Requirements as given in Section 5.4.3 are applicable.

ii. Load Factor Design Bolted Joints. Load factor design of eccentrically loaded joints is directly comparable to the allowable stress design. The design criteria provides that the load on a critical fastener at the factored load level does not exceed the shear capacity of the fastener multiplied by the reduction factor Φ. It follows that the factored joint loads $\bar{P}$ are given by

$$\bar{P} = \Phi CA_{SH}\tau_u \tag{13.23}$$

A Φ value of 0.75 was suggested for bolts in shear. This value of Φ is also applicable to eccentrically loaded joints. It yields shear stresses at the factored load level which are comparable to those obtained by factoring the allowable shear stresses by a factor of 1.75 or 1.85 for A325 bolts or A490 bolts, respectively.

DESIGN RECOMMENDATIONS ECCENTRICALLY LOADED JOINTS

Load Factor Design

Design strength for the fastener $\Phi\tau_u$ where

τ_u = average shear strength fastener = $0.60\sigma_u$
Φ = reduction factor = 0.75

Factored Joint Loads

$$P = \Phi C A_{SH} \tau_u$$

or

(a) if shear planes pass through bolt shank

$$P = \Phi C m A_b \tau_u$$

(b) if shear planes pass through the bolt threads

$$P = 0.75 \Phi C m A_b \tau_u$$

iii. Slip-Resistant Joints. For the analysis of slip-resistant joints it was assumed that at the slip load the maximum slip resistance R_S is reached for all fasteners. The slip resistance per fastener is equal to the product of the number of slip planes m, the nominal bolt shear area A_b, and the allowable shear stress τ_a. The allowable shear stress accounts for bolt quality, acceptable slip probability, tightening procedure, and fabrication factors. The recommendations given in Chapter 5 for slip-resistant shear splices are applicable to the design of eccentrically loaded slip-resistant joints as well.

Upon determining the allowable value of R_S the joint resistance can be evaluated from the analysis described in Section 13.3. The slip load P_{slip} can be determined from

$$P_{\text{slip}} = C' m \tau_a A_b \qquad (13.24)$$

where m defines the number of slip planes and C' is a factor which depends on the bolt pattern and the eccentricity of the load. An analysis of the slip load can be performed for different bolt patterns and eccentricities. Tables of C' values for fastener patterns comparable to those shown in Tables 13.1 and 13.2 have been developed for the case of uniform fastener resistance which is directly analogous to the slip condition. A comparison between the C' values given in Ref. 13.8 and the C values summarized in Tables 13.1 and 13.2 indicate only slight differences. The difference is generally less than 5%; for a few cases, the C' value exceeds C by about 10%. Considering these minor differences between the C and C' values, it is more convenient to consider a single set of coefficients for both types of joints. Hence the C values given in Tables 13.1 and 13.2 can be used to design slip-resistant joints as well. Since the C' value usually exceeds the C value slightly, this results in a conservative design.

Recently a pilot study was made to evaluate the load deformation behavior of eccentric slip-resistant joints.[13.12] Figure 13.7 summarizes the load

13.5 Design Recommendations

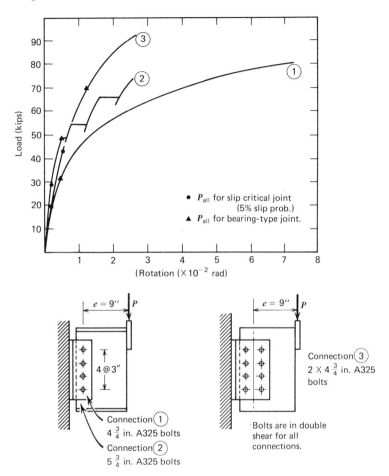

Fig. 13.7. Comparison of design recommendations and test data for slip-resistant eccentrically loaded joints.

rotation curves for three eccentrically loaded connections fastened by $3/4$-in. A325 bolts. Bolts were installed by the turn-of-the-nut method, and the faying surfaces were clean mill scale condition. The solid circular dots indicate the allowable slip load for each connection that results from the design recommendations given in this chapter for slip-resistant joints. A slip probability of 5% was selected, resulting in a 17.8 ksi allowable shear stress on the fastener (see Chapter 5.4). The allowable joint loads based on strength criteria only are indicated by the solid triangular shapes. A 30 ksi allowable shear stress was used for the fasteners.

DESIGN RECOMMENDATIONS SLIP-RESISTANT ECCENTRICALLY LOADED JOINTS

$$P_{\text{slip}} = CmA_b\tau_a$$

where τ_a = allowable shear stress for slip-resistant joint (see Section 5.4.2)
m = number of shear planes
A_b = nominal bolt area
C = coefficient as given in Tables 13.1 and 13.2 or determined from an analysis as discussed in this section.

To ensure minimum factor of safety with respect to the ultimate load of the connection:

$$A_b \tau_a \leq \tau_{\text{basic}} A_{\text{SH}}$$

where A_{SH} is the available shear area of the bolt.

References

13.1 C. Reilly, "Studies of Iron Girder Bridges," *Proceedings of the Institute Civil Engineers*, Vol. 29, 1870.
13.2 S. F. Crawford, and G. L. Kulak, "Eccentrically Loaded Bolted Connections," *Journal of the Structural Division, ASCE*, Vol. 97, No. ST3, March 1971.
13.3 E. Yarimci and R. G. Slutter, *Results of Tests on Riveted Connections*, Fritz Engineering Laboratory, Report 200.63.403.1, Bethlehem, Pa., April 1963.
13.4 T. R. Higgins, "New Formula for Fasteners Loaded Off Center," *Engineering News Record*, May 21, 1964.
13.5 C. L. Shermer, "Ultimate Strength Analysis and Design of Eccentrically Loaded Bolted or Riveted Fasteners," presented at the 1964 ASCE Annual Meeting and Structural Engineering Conference, New York, October 1964.
13.6 A. L. Abolitz, "Plastic Design of Eccentrically Loaded Fasteners," *Engineering Journal, AISC*, Vol. 3, No. 3, July 1966.
13.7 American Institute of Steel Construction, *Manual of Steel Construction*, 7th ed., New York, 1970.
13.8 C. L. Shermer, "Plastic Behavior of Eccentrically Loaded Connections," *Engineering Journal, AISC*, Vol. 8, No. 2, April 1972.
13.9 T. R. Higgins, "Treatment of Eccentrically-Loaded Connections in the AISC Manual," *Engineering Journal, AISC*, Vol. 8, No. 2, April 1971.
13.10 L. Tall (ed.), *Structural Steel Design*, Ronald Press, New York, 1964.
13.11 W. McGuire, *Steel Structures*, Prentice Hall, Englewood Cliffs, N.J., 1968.
13.12 G. L. Kulak, Private Communication, February 1972.

Chapter Fourteen

Combination Joints

14.1 INTRODUCTION

Most connections use a single fastening system to connect plates or members together and provide the means of transferring the forces acting in or on the joint. However, situations do arise where it is desirable or necessary to combine two methods of fastening in a connection. This generally involves rivets and bolts or bolts and welds. In these connections the two fastening systems share the load. Joints of this type are generally referred to as combination joints or load-sharing joints.

There are two general types of combination connections illustrated in Fig. 14.1. The one type, shown in Fig. 14.1a, utilizes two different fastening systems to share the load on a common shear plane. This condition may occur when reinforcing or strengthening an existing joint. For example, high-strength bolts may be used to replace several rivets. In other situations, space may not be available for additional fasteners and welds are added to the joint. In either case the applied loads are transferred by both types of fasteners on a common shear plane.

Combination joints which combine fasteners on a common shear plane have the advantage of being compact. This reduces the required space and the amount of splice material. In addition, they can help overcome field erection problems. Welded connections are generally more compact than bolted connections. However, fabrication tolerances for welding are more rigid than the tolerances allowed for bolted connections. Before the welding process is started, positioning and holding the components in place must also be considered and accounted for. Bolted connections with regular hole clearance ($\frac{1}{16}$-in.) provide for some relative movement between the connected parts after initial assembly and before final tightening of the bolts. Therefore, a member in a frame can be more easily installed with bolts. After the member has been positioned and aligned properly, the bolts are tightened. It is easy to add welds to a connection after it has been first bolted into place (see Fig. 14.1a).

Combination joints of the type as shown in Fig. 14.1a have a wide application for reinforcement of existing mechanically fastened joints. Simple

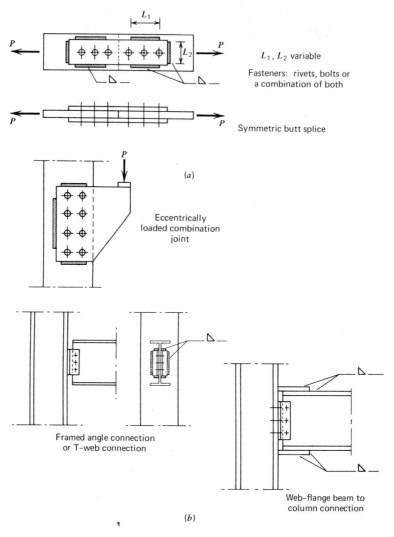

Fig. 14.1. Typical combination joints. (a) Load sharing on a common shear plane; (b) combination joints with two different shear planes.

14.2 Joints Which Share Load on a Common Shear Plane

shear splices or eccentrically loaded shear splices are typical connections that can utilize a combination of mechanical fasteners and welds on a common shear plane.

The behavior of small combination joints with bolts and welds or with bolts and rivets combined on a single shear plane has been studied to evaluate joint behavior and develop design recommendations.[5.5, 9.2, 14.1, 14.2] These tests have demonstrated the applicability of this type of joint. The work in this area is not extensive and further research would be desirable.

In the other major type of combination connections two different fastening methods are used but not on a common shear plane. Examples of this category of combination joints are as shown in Fig. 14.1b. These connections include the simple combination framed beam connection which utilizes shop welds to connect the web angles to either the beam web or the member the beam frames into, and bolts for the field connection. In this particular case, both the bolts and the welds are resisting the beam shear force. Other variations of this type of combination joint are possible such as welding the flanges of beam to column joints and providing a bolted shear connection for the web.

Usually this type of combination joint will provide greater economy and allow increased flexibility during erection. Many possibilities for combination joints exist which will only depend on the ingenuity of the engineer. All available evidence shows that they provide a satisfactory joint with adequate strength and stiffness when proper design procedures are used for the component parts.[14.4]

The remainder of this chapter discusses the behavior of bolted-welded and riveted-bolted-type combination joints where the fasteners are sharing the load on common shear plane. Other combinations of fastening systems are not considered for this type of combination joint because of the lack of information and because of their limited use in structural applications.

Discussion of the behavior of the other major type of combination connections where different types of fasteners are used, but not on a common shear plane, is given in Chapter 18.

14.2 BEHAVIOR OF COMBINATION JOINTS WHICH SHARE LOAD ON A COMMON SHEAR PLANE

Before the combined action of two different fastening methods acting in a common shear plane is discussed, it is desirable to reexamine the load-deformation behavior of the different types of individual fasteners. Figure 14.2 shows typical load-deformation curves for welded, bolted, and riveted tension specimens. This figure indicates that high-strength bolted connections with normal hole clearance provide a very high initial stiffness up to

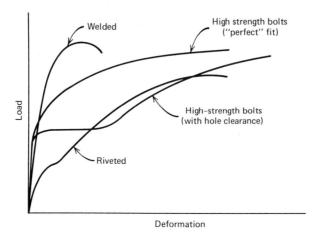

Fig. 14.2. Load-deformation relationships for different fastening methods (Ref. 9.2).

the slip load of the connection. During slip, the deformations increase significantly until the bolts come into bearing. After the bolts are in bearing, the load-deformation curve shows an increase in joint stiffness. Joint slip can be minimized by installing fitted bolts in matching drilled holes.

Compared to slip-resistant high-strength bolted joints where the load is transferred by friction, riveted connections are generally more flexible. Often a sudden change in the slope of the load deflection curve can be observed which is directly comparable to slip in a high-strength bolted connection. This "slip" is usually less than one third the slip observed in high-strength bolted connections.

A typical characteristic of a welded connection as compared to riveted or high-strength bolted connections is the reduced deformation capacity of the shear connection. Slip does not occur in welded connections and the initial stiffness of the joint only changes as the ultimate load is approached. From these load-deformation relationships for typical fasteners, one can conclude that combination of these fasteners would be most appropriate where compatible deformation characteristics exist. This appears to be with welds and slip-resistant high-strength bolts or with rivets and bolts.

14.2.1 High-Strength Bolts Combined with Welds

A comparison of the load-deformation capacity of welded and high-strength bolted connections with normal $\frac{1}{16}$-in.-hole clearance indicates that the total deformation capacity of the welds is about the same order of magnitude as the maximum slip of a high-strength bolted connection.

14.2 Joints Which Share Load on a Common Shear Plane

Therefore, if both fastening methods are used on a common shear plane, the capacity of the resulting combination joint might be approximated by summing the strengths of the welds and the slip resistance of the bolted connection. Tests have been performed to evaluate the validity of this estimate of the ultimate strength of bolted-welded combination joints.[5.5, 9.2, 14.1, 14.2] The test joints were generally small tension type butt splices with two bolts on either side of the splice, as shown in Fig. 14.3. The influence of the location of the welds, that is, either transverse or parallel

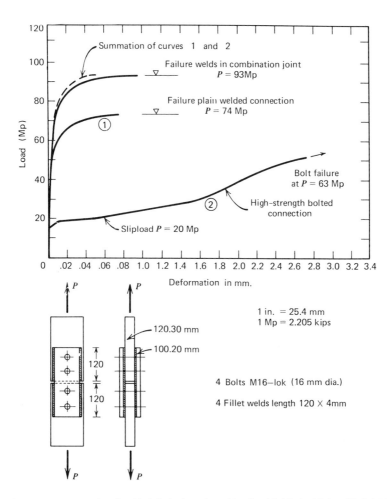

Fig. 14.3. Test results of welded, bolted, and combined welded-bolted joints (Ref. 9.2).

to the applied load, was also studied. Furthermore, the ratio of the capacity of the welds with respect to the slip resistance of the bolts was considered as a test variable.

Figure 14.3 summarizes the results observed in a typical series of test joints.[9.2] The load-deformation behavior of the plain welded and the plain bolted connection is shown as well as the load-deformation behavior of the combination bolted and welded joint. It is apparent that the behavior of the combination joint can be adequately approximated on the basis of the behavior of the welds and bolts alone. Furthermore, these results indicate that the capacity of the combination joint is provided by the sum of the slip load of the plain bolted connection and the strength of the welds. Other combinations of weld length, weld location, and slip resistance of the bolted joint resulted in similar conclusions.[9.2]

The tests reported in Ref. 9.2 were limited to small connections with only a few bolts in line. In larger connections some misalignment may exist and the bolts come into bearing before failure of welds occurs. The load carried by the bolted connection is then transmitted by friction and bearing. The failure load of these connections is likely to exceed the estimated ultimate load determined from the slip resistance of the bolts and the strength of the welds. Reducing the hole clearance would also bring the bolts into bearing and increase the ultimate strength of a bolted-welded combination joint. The maximum capacity of a combination joint is developed when fitted bolts are installed in matching drilled holes. Tests have indicated that these connections have an ultimate load that exceeds the summation of the weld strength and the slip load of the bolted connection.[9.2, 14.2] Obviously such joints are not very economical.

Another aspect that has to be considered is the behavior of combination joints under repeated loading conditions. The behavior of high-strength bolted connections subjected to repeated loading conditions is discussed in Chapter 5. Tests performed in Germany indicated that the fatigue strength of a high-strength bolted connection decreases when weldments are added.[9.2] This reduction in fatigue strength is expected because the weld toe is the critical region and crack growth will occur just as in a welded joint. The weld toe was more critical than the bolt holes in all test joints.[9.2] A comparison of the few data available with welded joint data indicates that the fatigue strength is not significantly different from the fatigue strength of a similar plain fillet welded connection. Hence the design criteria for welded joints should be used for cyclic load conditions when the welds are positioned on the boundaries of the combination joint.

Recent tests have indicated that an improvement in fatigue strength can result when the welds are placed on the joint interior.[14.3] This removes the

14.2 Joints Which Share Load on a Common Shear Plane

weld from the more highly stressed joint boundary where the geometric discontinuity is more severe and places it in a lower stressed region. In addition, the stress concentration condition is generally decreased, since the connected parts are more nearly subjected to about the same strain conditions. The research available is not extensive enough to develop design criteria at this time.

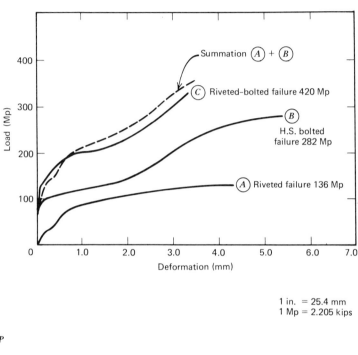

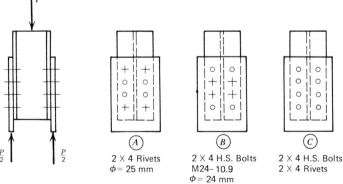

Fig. 14.4. Test results of riveted, bolted, and combined riveted-bolted joints (Ref. 9.2).

14.2.2 High-Strength Bolts Combined with Rivets

A combination of rivets and high-strength bolts intersecting the same shear plane would not be used with new construction. However, high-strength bolts are often used to replace one or more rivets in existing riveted connections. This is done to either repair the joint or to strengthen the connection.

Combining high-strength bolts and rivets in one connection has several advantages. The rivets have less hole clearance which decreases the slip occurring when the slip resistance is exceeded for the high-strength bolted connection. In addition, the high-strength bolts increase the connection stiffness when their slip resistance is not exceeded. Furthermore, replacing rivets by high-strength bolts generally improves the fatigue strength as well as the static strength of a riveted connection significantly.[9.2]

Tests to evaluate the load deformation behavior of short bolted-riveted combination joints have indicated that the ultimate strength of the joint is adequately approximated by the summation of the resistance of the two types of fasteners. This is illustrated in Fig. 14.4 where the load-deformation curves of a riveted, a bolted, and a bolted-riveted combination joint are compared. This figure clearly shows the increased stiffness of the combined joint as compared to the riveted joint. The improved slip behavior of the combination joint is also evident.

Since the joint strength of short combination joints is an aggregate of the strengths of the individual fasteners, it does not matter how the fasteners are arranged in the combination joint. Hence either the outermost rivets or rivets located in the joint interior can be replaced by high-strength bolts. Both joints yield about the same ultimate load. Based upon the observed behavior of long riveted and bolted joints, the fastener location will influence the joint strength. Because of "unbuttoning," replacing the outermost rivets of a long joint by high-strength bolts will be more effective in increasing the joint strength than replacing the same number of interior fasteners. Experimental verification is not available on long joints at the present time (1973).

Many test programs have indicated that high-strength bolted shear splices subjected to repeated type loading generally exhibit a significant higher number of load cycles before failure than comparable riveted specimens (see Chapter 5). This difference is mainly attributed to the high clamping force provided by the bolts, which results in a more favorable stress distribution around the bolt hole as compared to the stress flow around the holes in a riveted connection. Hence the replacement of rivets by high-strength bolts will increase the fatigue strength of a connection. Tests on small bolted-riveted combination joints have confirmed this con-

14.3 Design Recommendations

clusion.[9.2] Since the end of a connection is more critical because the stresses in the connected plates are higher, replacing the outer rivets by high-strength bolts is the only effective way to increase the fatigue strength of a riveted connection.

14.3 DESIGN RECOMMENDATIONS

Although only limited test data are available, a knowledge of the behavior of the different fastener methods enables design recommendations to be developed for combination joints which utilize two different types of fasteners to transfer load on a common shear plane.

14.3.1 Static Loading Conditions

The ultimate load of a bolted-welded combination joint can be estimated as the sum of the slip resistance of the bolted parts and the ultimate load of the plain welded connection. The stiffness of a welded-bolted combination joint is comparable to the stiffness of a slip-resistant high-strength bolted connection.

The allowable load of a bolted-welded combination joint consists of two contributions. One contribution results from the slip resistance of the bolted parts. The second contribution results from the resistance of the welds. The allowable load of the welded connection can be determined on the basis of the allowable stresses given by applicable specifications. The contribution of the slip-resistant bolted parts can be evaluated on the basis of the design recommendations given in Chapter 5. This takes into account such factors as faying surface condition, bolt grade, and tightening procedures.

The allowable load on a riveted-bolted combination joint is equal to the sum of the allowable loads on the individual fasteners when they share a common shear plane. Both slip-resistant and bearing-type connections are applicable. The allowable load component resisted by the high-strength bolts in the connection can be based on the allowable shear stress as given in Chapter 5 for either slip-resistant or bearing-type connections.

14.3.2 Repeated Loading

When high-strength bolts and fillet welds are combined to resist forces on a common shear plane, the fatigue strength is governed by the welded joint when the welds are placed on the exterior of the joint. Crack growth occurs first from the weld toe termination and fatigue provisions for the welded detail should be used for design.

An improvement in fatigue strength has been observed when the welds are placed in the interior of the joint in a less highly stressed region.[14.3]

Data available are not sufficient to develop general design recommendations at this time.

When high-strength bolts have been used to strengthen riveted joints, an improvement in fatigue strength has been noted when the bolts were placed at the joint ends where the stressed plates are most critical. However, data are not sufficiently comprehensive to develop design criteria and take advantage of the observed increase. Therefore, it is recommended that the fatigue design provisions in use for riveted connections be used for combined joints with rivets and bolts resisting forces on a common shear plane.

References

14.1 W. Hoyer and H. Skwirblies, *Hochfeste Schrauben in Verbindungen Mit Schweissnachten*, (2nd report), Wissenschaftliches Zeitschrift der Hochschule fuer Bauwesen, Cottbus, 1959/1960, Vol. 1, Cottbus, Germany, 1960.

14.2 N. M. Holz and G. L. Kulak, *High Strength Bolts and Welds in Load-Sharing Systems*, Department of Civil Engineering, Nova Scotia Technical College, Nova Scotia, September 1970.

14.3 E. Ypeij, *New Developments in Dutch Steel Bridge Buildings*, Preliminary Report 9th Congress IABSE, Amsterdam, May 1972.

14.4 J. S. Huang, W. F. Chen, and J. E. Regec, *Test Program of Steel Beam-to-Column Connections*, Fritz Engineering Laboratory Report 333.15, Lehigh University, Bethlehem, Pa., July 1971.

Chapter Fifteen
Gusset Plates

15.1 INTRODUCTION

If two or more members join at a section and their centroidal axes do not coincide, gusset plates are used to transfer the forces as illustrated in Fig. 2.4c. Splice plates in butt joints are only subjected to tensile or compressive forces, whereas gusset plates are usually subjected to bending, shear, and axial force components as a result of the loads in the members. The forces enter and emerge from gusset plates by shear transfer through mechanical fasteners or weldments.

Out-of-plane bending in gusset plates is generally insignificant. Often the load application is symmetric with respect to the plane of the gusset plate, or joint geometry prevents or minimizes the secondary out-of-plane bending stresses as shown in Fig. 2.4c. Because of these factors, gusset plates generally are treated as two-dimensional plane stress problems. Secondary stresses due to out-of-plane bending are neglected in design.

Comparatively few attempts have been made to determine the stress distribution in gusset plates. The current (1973) design of gussets is largely the result of experience, general practice, and intuition on the part of the designer. The available experimental and theoretical work has concentrated largely on an elastic analysis.[15.1, 15.2, 15.5] Design rules have been developed from these studies, but no adequate assessment of the ultimate strength and stress distributions has been attained. Recently (1972), the finite element method has been applied to this type of problem. This has made it possible to evaluate the stress distribution in gusset plates at the various load stages in the elastic [15.4, 15.7, 15.8] as well as in the inelastic range.[15.8] The ultimate strength of gusset plates was also predicted.[15.8] It appears desirable to develop further experimental data on the behavior of gusset plates at various load stages including its ultimate load. This would permit the analytical studies to be evaluated for reliability before more extensive evaluations were made. A more rational design of gusset plates may result from these studies.

This chapter discusses the methods currently in use for the design of gusset plates. An examination of current practice suggests that substantial variations in the factor of safety against ultimate load exists in gusset plates because of the assumptions involved. Despite these shortcomings of the presently available design methods, these procedures continue to be used because experience with these methods has resulted in gusset plates that have provided satisfactory performance and behavior. There are, no doubt, substantial variations in the actual strength of the various gusset plates that result from this design approach. However, there are no known failures or documented cases of adverse behavior.

15.2 METHOD OF ANALYSIS AND EXPERIMENTAL WORK ON GUSSET PLATES

The design of gusset plates has long been based on simple methods of analysis. Simple strength of materials analysis or specification rules were used.[15.5] Such an analysis is based on assumptions, and their adequacy is not fully known.

The procedure generally followed and presented in many design handbooks is summarized as follows.[15.6] It is assumed that all fasteners connecting a member to the gusset carry an equal share of the load. This permits the number of fasteners required to transmit the load from each member into the gusset plate to be determined. Note that comparable assumptions regarding the load transfer are used for design of other types of shear splices. The planar dimensions of the plate are selected so that all fasteners can be placed. A tentative plate thickness is selected, often on the basis of

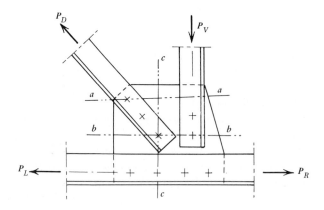

Fig. 15.1. Analysis of gusset plates. Bending stress $\sigma_{max} = P/A \pm Mc/I$. Shear stress $\tau_{max} = \frac{3}{2} V/A$. a-a, b-b, c-c denote sections to be checked.

15.2 Method of Analysis and Experimental Work on Gusset Plates

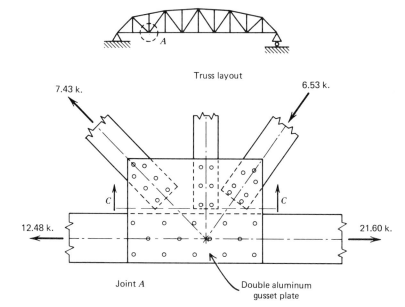

Fig. 15.2. Gusset plate model as used by Whitmore (Ref. 15.2).

experience of the designer or as prescribed by applicable specifications. Stresses are then evaluated on the most critical section by assuming the plate to act as a beam. Hence beam theory is used to evaluate the stresses at the selected section. Generally the analysis consists of checking various sections through the plate in order to obtain the governing one (see Fig. 15.1).

It has been recognized for long that the beam method of analysis is of questionable value.[15.1-15.6] The load partition among fasteners connecting a member to a gusset plate is generally not uniform, furthermore, the applicability of beam formulas to the geometries generally encountered in gusset plates is questionable. To examine the validity of the use of beam formulas for this problem, Whitmore, in 1952, investigated the stress distribution in a 12.6 × ⅛ × 16.6 in. aluminum gusset plate in which the connections were made by tight fitting pins and bolts.[15.2] The model simulated a lower chord joint of a Warren-type truss with a continuous chord (see Fig. 15.2). A vertical member was attached to the model but not loaded. Whitmore observed that the locations of the maximum tensile and compressive stress were near the ends of the tension and compression diagonals, respectively. The assumption that normal stresses, bending stresses, and shear stresses

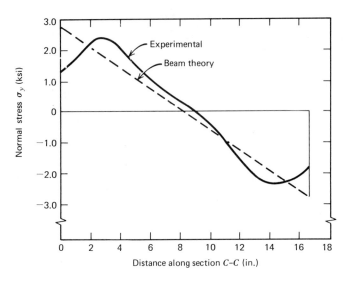

Fig. 15.3. Distribution of vertical normal stress on critical section C–C: see Fig. 15.2.

on a critical plane through the ends of the diagonals are distributed according to beam formulas was found to be inaccurate. This is illustrated in Fig. 15.3 where the distribution of the vertical normal stress along a section parallel to the chord member and passing through each diagonal is shown. A significant difference beween the calculated and observed stresses is noted, particularly at the edges of the plate.

Whitmore concluded that the maximum normal stress at the end of a member could be estimated adequately by assuming that the member force was distributed uniformly over an effective area of plate material. This area was obtained by multiplying the thickness of the plate by an effective length. The effective length was estimated by constructing 30° lines from the outer fasteners in the first row to their intersection with a line perpendicular to the line of action of the external load and passing through the bottom row of fasteners, as shown in Fig. 15.4. The line segment intercepted by the 30° line is then used as the effective width of the plate.

Experimental information about the stress distribution in gusset plates is scarce, perhaps because of the difficulties involved.[15.4] Relatively few publications have treated the subject in recent years.[15.1-15.4] Except for the effective width solution proposed by Whitmore, no other design recommendations have been suggested in the literature.

Methods of analysis have become available, such as the finite element method, which permit the gusset plate to be analyzed in the elastic and

15.2 Method of Analysis and Experimental Work on Gusset Plates

inelastic ranges. Vasarhelyi[15.4] and Davis[15.7] both attempted an elastic finite element solution of specific gusset plates. Struik[15.8] not only studied the problem in the elastic range but also predicted the behavior of gusset plates in the inelastic range up to their ultimate strength. In the elastic-plastic analysis the presence of the holes was accounted for in an approximate manner.

The elastic analyses[15.4, 15.7, 15.8] confirmed Whitmore's conclusions. Significant variation between stress distributions predicted by the finite element method and beam theory existed. However, the difference was not necessarily unsafe. None of the stresses evaluated by the finite element analyses exceeded the maximum values predicted by beam theory. The location and distribution of the maximum stresses showed substantial variation.

Some of the results of the elastic-plastic finite element analysis[15.8] of a typical gusset plate are shown in Figs. 15.5 through 15.7. Figure 15.5 shows the geometry of the gusset plate as well as the applied loads. The tensile strength of the material was assumed to be 70 ksi at a strain of 15%. Reaching the tensile strength in one or more elements was considered to result in failure of the gusset and defined the ultimate load.

The predicted load-displacement curves for two typical points on the gusset are shown in Fig. 15.6. The elastic-plastic boundaries corresponding to the load levels P_1, P_2, and P_3, indicated in Fig. 15.6, are summarized in Fig. 15.7. It is apparent that yielding occurred near the ends of the members soon after load P_1 was applied. The load deformation curves start to deviate from linearity, reflecting plastification of the section. At load stage P_3 the system exhibited substantial nonlinear behavior. The tensile strength was first reached in the elements at the end of the diagonal members, as indicated in Fig. 15.6.

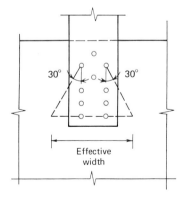

Fig. 15.4. Evaluation of effective width for fastener pattern.

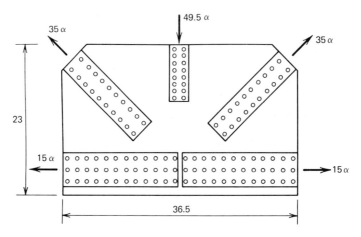

Fig. 15.5. Geometry and loading conditions for sample gusset plate. Fastener holes 0.5 in dia. Plate thickness 0.25 in. α Load parameter to indicate proportional loading (see Fig. 15.7).

The allowable loads for this particular gusset plate were evaluated on the basis of the current AISC specifications[2,11] and are also shown in Fig. 15.6. The elastic plastic analysis indicated a factor of safety against ultimate between 2.5 and 2.7, depending on the method of analysis used. For this particular example the 30° effective width method gave a slightly higher allowable load than beam theory. On the basis of these finite element studies, it was concluded that current design procedures result in a variable factor of safety against the gusset plate capacity.[15.8]

15.3 DESIGN RECOMMENDATIONS

Design recommendations for gusseted connections concern the fasteners as well as the plate material. To determine the total number of fasteners required to transfer the load from a member into the gusset plate, equal load distribution among the fasteners may be assumed, as is done with other joints. Design recommendations for fasteners are given in Chapter 5 for symmetric butt splices and are applicable to the design of slip resistant and bearing-type gusset plates as well.

The analysis of the gusset plate can be performed by evaluating the critical normal stress as was recommended by Whitmore in Ref. 15.2. This method requires the evaluation of an effective plate area as indicated in Fig. 15.4. The normal stress on this effective area should not exceed the allowable stress permitted by the appropriate specification. Bearing stresses on the plate material must be within the limitations provided in Chapter 5.

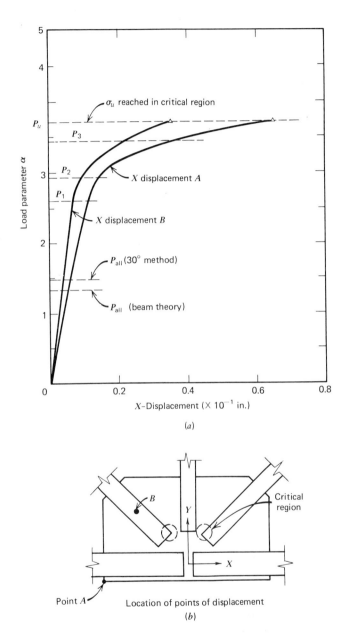

Fig. 15.6. Typical load displacement curves for sample gusset plate. (a) Load-displacement curves; (b) location of points of displacement.

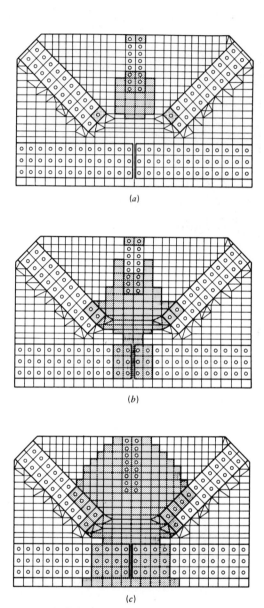

Fig. 15.7. Elastic plastic boundary at various load stages (for load reference see Fig. 15.6). (a) Load stage P_1; (b) load stage P_2; (c) load stage P_3.

The effective width method guards against a tearing or crushing-type failure in the gusset plate near sections where members are terminated. As an alternative solution the beam formulas can be applied to the critical section and maximum stresses evaluated. Although some questions are raised as to the applicability of these formulas, past practice has shown that this method results in an adequate and safe design. Until further analytical and experimental work is available, a more rational design method cannot be developed nor an estimate of the ultimate capacity given.

References

15.1 T. H. Rust, "Specification and Design of Steel Gusset Plates," *Transactions, ASCE*, Vol. 105, 1940.

15.2 R. E. Whitmore, *Experimental Investigation of Stresses in Gusset Plates*, University of Tennessee Engineering Experiment Station Bulletin 16, May 1952.

15.3 P. C. Birkemoe, R. A. Eubanks, and W. H. Munse, *Distribution of Stresses and Partition of Loads in Gusseted Connections*, Structural Research Series Report 343, Department of Civil Engineering, University of Illinois, Urbana, March 1969.

15.4 D. D.Vasarhelyi, "Tests of Gusset Plate Models," *Journal of the Structural Division, ASCE*, Vol. 97, ST2, February 1971.

15.5 J. A. L. Waddell, *Bridge Engineering*, Wiley, New York, 1916.

15.6 E. H. Gaylord and C. N. Gaylord, *Design of Steel Structures*, McGraw-Hill, New York, 1957.

15.7 C. S. Davis, "Computer Analysis of the Stresses in a Gusset Plate," M. S. Thesis, Department of Civil Engineering, University of Washington, Seattle, 1967.

15.8 J. H. A. Struik, "Applications of Finite Element Analysis to Non-Linear Plane Stress Problems," Ph. D. Dissertation, Department of Civil Engineering, Lehigh University, Bethlehem, Pa., November 1972.

Chapter Sixteen

Beam and Girder Splices

16.1 INTRODUCTION

Splices in beams and girders are generally classified either as shop or field splices. Shop splices are made during the fabrication of the member in the shop. They are usually required to overcome length limitations of structural components as a result of fabrication or transportation facilities. The location of a shop splice in a member is often determined by loading conditions or stress resultants acting on the member and by the available lengths of material.

Field splices are necessary when a structural member becomes too long to be transported in one piece from the shop to the construction site. Occasionally, the available equipment in the field may also limit the maximum size or weight of structural components. Such limitations may require additional field splices.

This chapter deals specifically with the analysis and design of bolted or riveted beam and girder splices. Current practice varies and is largely based on past experience and limited experiment data.[16.1-16.3] Most designs involve equilibrium checks of the joint components. The stresses are computed on the basis of an assumed elastic behavior of all the structural components. Past practice has shown that this procedure results in a satisfactory design when the connection is subjected to static loading. Further work may lead to the development of more rational methods of analysis for this type of splice.

16.2 TYPES AND BEHAVIOR OF BEAM-GIRDER SPLICES

Two types of connections are currently in use for bolted beam-girder splices. They are (a) the end-plate connection, and (b) the more commonly used web-flange splice. Both connections are shown in Fig. 16.1. The major difference between these two types of joints is the loading condition to which the fasteners are subjected. The fasteners in the end-plate connection are generally subjected to a combined axial force and shear force, whereas the fasteners in the web-flange-type splice are subjected to shear alone. The

16.2 Types and Behavior of Beam-Girder Splices

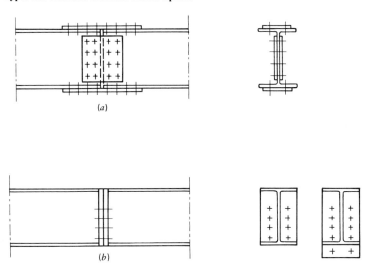

Fig. 16.1. Beam-girder splices. (a) Web-flange splice; (b) end plate splice.

end-plate connection is also used as a moment resistant beam-to-column connection. Design recommendations for beam-to-column joints are discussed in Chapter 18. In this chapter emphasis is placed on the design of web-flange-type splices.

Usually two bolts are placed in the compression region of an end-plate connection. Although these bolts do not actively participate in transferring the moment, they are desirable from a practical point of view and hold the joint together. They also increase the shear capacity of the joint. In addition to the bolts in the compression region, a cluster of bolts is placed near the tension flange to obtain the maximum moment resistance for a given number of bolts and type of end plate. The fasteners near the tension flange can be used even more effectively if the end plate is extended beyond the tension flange and bolts are placed in this region as well (see Fig. 16.1).

As a moment connection the end-plate splice is most economical in relatively light constructional steelwork because it requires less material and fasteners than conventional web-flange splices. Satisfactory behavior up to the plastic limit load of the beam can be achieved if the fasteners are adequately designed. This is illustrated in Fig. 16.2 where load versus midspan deflection curves are compared for beams with two types of end-plate splices in the constant moment region.[16.2] The observed behavior is almost identical to the behavior of plain beams. The plastic moment, M_p, for the gross section of the beam was reached and sustained.

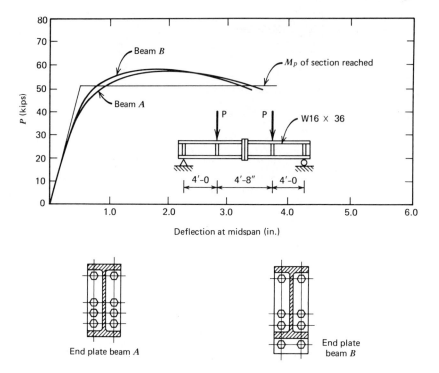

Fig. 16.2. Load deflection diagrams for beams with end plate splice (Ref. 16.2).

As beam sizes are increased or large shear forces are to be transferred, the end-plate splice looses much of its economy and is replaced by the conventional beam splice shown in Fig. 16.1a. The location of the web and flange splices may be staggered, but this is often avoided to simplify field assembly.

Current analysis of girder splices assumes that the web transmits the shear force and the flange splices resist the moment. Although the validity of these assumptions has not been extensively verified by experiments, past experience and available test results indicate that the assumptions are reasonable. Hence the analysis and design of a girder splice can be divided into two parts (1) resistance of the flange splices to the applied moments, and (2) the shear resistance of the web splice.

16.2.1. Flange Splices

Investigations were performed to determine the ultimate resisting moment of a beam with fastener holes in both flanges.[16.1-16.3] The general objective of

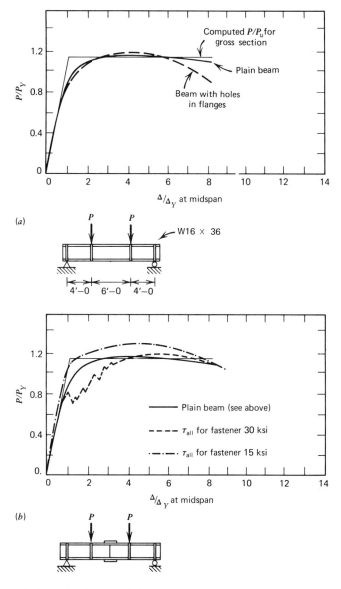

Fig. 16.3. Typical load-deflection curves (Ref. 16.2).

these investigations was to evaluate whether the gross section plastic moment could be developed and whether the connection could provide sufficient rotation capacity. An extensive test series was reported in Ref. 16.2. Plain beams, beams with holes in the flanges, and beams with a flange splice in the constant moment region were tested. Single splice plates were bolted on the outside of the flanges and the allowable fastener shear varied from 15 to 30 ksi. Typical results are shown in Fig. 16.3. The nondimensional load-deflection curves show the ratio of load to first yield load and deflection to yield load deflection. Figure 16.3a compares the behavior of a plain rolled beam to that of a beam with holes in the flanges. No splice plates were provided. It is visually apparent that the holes did not affect the beam capacity. Figure 16.3b shows the load-deformation behavior of similar beams with the flanges spliced. The required number of fasteners for the splice was based on an allowable shear stress for the fasteners of 15 ksi for one beam, which resulted in 48 fasteners per splice. A second beam was designed using 30 ksi in shear which resulted in 24 fasteners per splice. An allowable shear stress of 15 ksi for clean mill scale surfaces is a conservative estimate of the capacity of a slip-resistant joint. Therefore, slip was not expected to develop in this joint and did not occur.

All tests developed the gross section plastic moment even though two $^{15}/_{16}$-in.-diameter holes were placed in each flange cross-section. This reduced the flange area by 23%. Nevertheless, the beams were all able to develop the full plastic moment of the gross section. The holes in the flanges did not decrease the moment capacity of the beams.[16.2] The holes only influence the strain in the flanges locally. The material near the net section at the holes strain-hardens and permitted the full plastic moment of the gross section to be reached. This behavior of the net section is related to the ratio of the net to gross section area of the flanges, as was noted in Chapter 5.

Figure 16.3b shows that the slip between the splice plates and the flanges influences the load-deformation behavior of the beam but has a negligible effect on the ultimate moment capacity of the beam.[16.2, 16.3] At ultimate, plastic hinges formed in the constant moment region and failure occurred by local buckling of the compression flange.

The beam tests reported in Ref. 16.2 only used flange splices in the constant moment region and no web splice was present. Beams spliced in the constant moment region with both web and flange splice plates are reported in Ref. 16.1. The observed maximum moment capacity was approximately equal to the gross section plastic moment. Hence providing web splice plates did not significantly alter the moment capacity of the beam. Flange splices alone can be assumed to transfer the moment.

16.3 Design Recommendations

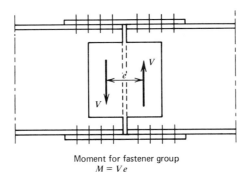

Moment for fastener group
$M = Ve$

Fig. 16.4. Design condition for fastener group in web splice.

16.2.2 Web Splices

Current design of web splices assumes that the web components primarily transfer the shear, an assumption which is based on beam theory and experimental observations. Sometimes the portion of the moment carried by the web is assumed to be transmitted by the web splice as well. In most situations, the moment carried by the web of a beam or girder is relatively small and is not considered in the design of the web splices.

The fastener group is subjected to an eccentrically applied shear force as illustrated in Fig. 16.4. Since the splice plates transfer shear forces across the discontinuous web, the fasteners must be designed to reflect the eccentric shear force shown in Fig. 16.4. The amount of eccentricity of the shear force may be conservatively taken as the distance between the centroids of the fastener groups on either side of the splice. Existing experimental data do not provide much insight into the exact distribution of force in the web shear splice nor has this force been systematically studied. The resultant shear force may actually provide a less severe condition of eccentricity than assumed. However, test data are needed before a reduced eccentricity can be considered.

16.3 DESIGN RECOMMENDATIONS

16.3.1 Flange Splices

The fasteners in the flanges must resist the force, M/d. A single shear splice plate on each flange is often sufficient. For large shapes and heavy flanges, splice plates may be required on both sides of the flanges to reduce the number of fasteners by providing a double shear condition and to reduce the splice plate thickness. The fasteners can be designed using the recommendations given in Chapter 5 for symmetric butt joints. Depending on the required joint performance, slip-resistant as well as bearing-type joints can be used.

The moment capacity of the beam is not affected by the reduction in cross-sectional area caused by the fastener holes unless the ratio of net section to gross section area of the flanges (the A_n/A_g ratio) is less than $\sigma_y/0.85\sigma_u$ (see Chapter 5). The flange splice plates in the tension region should be treated as tension members and are also subject to the design recommendations given in Chapter 5.

16.3.2 Web Splices

The fasteners in the web splice should be designed for the eccentric shear force at the spliced section. On the basis of available data, the eccentricity e is conservatively estimated as the distance between the centroids of the fastener groups on each side of the splice. With this defined loading condition the design recommendations given in Chapter 13 for eccentrically loaded joints are applicable to these joint conditions as well. Slip-resistant as well as bearing-type joints can be designed, and the choice only depends on the required joint performance.

Two web splice plates, one on either side of the web, are recommended for beam or girder splices. This not only creates a symmetric load transfer with respect to the plane of the web, but also the fasteners are subjected to double shear conditions which reduces the required number of fasteners and thus the eccentricity.

The overall dimensions of the web splice plates depend on the selected fastener pattern. The thickness of the splice plate can be determined from the applied eccentric shear load and the applicable shear, bending, and bearing stresses.

The fastener shear stresses and the bearing stresses suggested in Chapter 5 were shown in Ref. 18.7 to be fully applicable.

References

16.1 F. W. Schutz, Jr., "Strength of Moment Connections Using High Tensile Strength Bolts," *AISC National Engineering Conference, Proceedings*, 1959.

16.2 R. T. Douty, and W. McGuire, "High Strength Bolted Moment Connections," *Journal of the Structural Division, ASCE*, Vol. 91, ST2, April 1965.

16.3 L. G. Johnson, J. C. Cannon, and L. A. Spooner, "Joints in High Tensile Preloaded Bolts—Tests on Joints Designed to Develop Full Plastic Moments on Connected Members," Jubilee Symposium on High Strength Bolts, Institution of Structural Engineers, London, June 1959.

Chapter Seventeen
Tension-Type Connections

17.1 INTRODUCTION

Fasteners are often subjected to a tensile-type loading by T-stubs or their equivalent. Some typical examples in this category are the hanger connection, the diagonal brace connection, and the structural beam-to-column connections shown in Fig. 17.1. Depending on the direction of the bending moment, either the top or bottom flange T-stub in a beam-to-column connection (Fig. 17.1a) is stressed in tension. It has long been recognized that excessive deformation of the T-stub results in additional fastener tension.[17.1] This phenomenon is called prying action. Tests have indicated that prying action reduces both the ultimate load capacity as well as the fatigue strength of bolted and riveted joints.[16.2, 17.1-17.4]

17.2 SINGLE FASTENERS IN TENSION

Cooling of hot driven rivets as well as tightening of a nut on a bolt results in an axial force or preload in the fastener. Inasmuch as this stress exists prior to the application of external loading, the fastener is prestressed. As a result of this preload, the external applied loads mainly change the contact pressure between the plates. Very little additional fastener elongation is introduced; hence there is only a minor change in bolt tension. This behavior can be illustrated by the model shown in Fig. 17.2.[13.11, 17.7] Tightening of the nut results in a tension in the bolt and compression between the two plates. Assuming that the bolts and plates remain elastic, the force in each is proportional to its change in length or

$$\Delta B = k_b \Delta e$$

and

$$\Delta C = -k_p \Delta e$$

where B represents the bolt preload, C the summation of contact forces between the plates, and k_b and k_p the stiffness of the bolt and the gripped plates, respectively. Also, Δe represents the change in bolt elongation due to an external applied load. This change in bolt elongation is also equal to

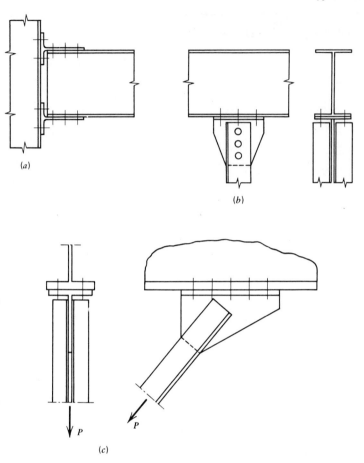

Fig. 17.1. Typical uses of T-type structural connections. (*a*) Beam-to-column connection; (*b*) hanger connection; (*c*) diagonal brace connection.

the expansion of the precompressed plates as long as separation of the plates does not occur. For the usual bolt and plate combinations k_p will be much larger than k_b, because the force B_0 is concentrated in the bolt whereas the force C_i is distributed over a much larger effective area of the plates. If no load is applied to the connection, the bolt preload B_0 and the contact forces C_i are equal. When a load T is applied to the outer surfaces of the plate, the fastener will elongate and the precompressed plates tend to expand to their original thickness. If the expansion does not exceed the initial contraction of the plates (see Fig. 17.2) some contact pressure will

17.2 Single Fasteners in Tension

remain. Equilibrium requires that

$$B = C_p + T \tag{17.3}$$

where T is the external applied load, C_p the summation of the reduced contact forces, and B the bolt force under an applied force T. Under such conditions an increase in applied load T results in an increase in bolt elongation Δe. The plates expand the same amount Δe. Because of the differences in stiffness of bolt and plates, Fig. 17.2 illustrates that the addition of an external force T results in a greater change in the compression in the plates (depicted as ΔC) than in the tension in the bolt, indicated as ΔB. A further increase in the external load T results in a decrease in the plate contact pressure C_p until the plates separate. For elastic conditions separa-

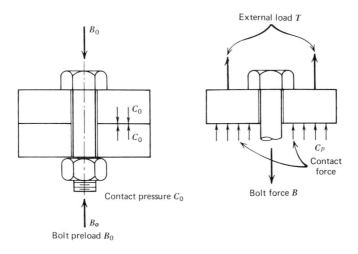

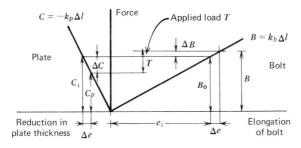

Fig. 17.2. Force in prestressed fastener.

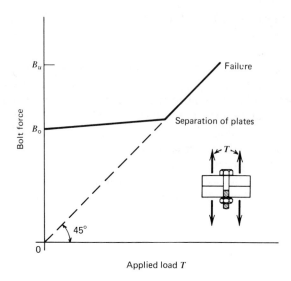

Fig. 17.3. Bolt force versus applied load for prestressed single bolt connection.

tion of the plates takes place at an applied load equal to

$$T = B_0 \left[1 + \frac{k_b}{k_p}\right] \tag{17.4}$$

After the plates are separated, the bolt force B is equal to the external applied load T.

The complete variation of the bolt force as a function of the applied load is given in Fig. 17.3. The factor k_b/k_p depends on actual dimensions of the connection. However, for most practical cases the ratio varies between 0.05 and 0.10. Hence unless separation of the plates takes place, the maximum increase in bolt force due to an applied external load is of the order of 5 to 10% of the initial bolt preload.

17.3 BOLT GROUPS LOADED IN TENSION—PRYING ACTION

One of the simplest connections with the bolt groups in tension is the symmetric T-stub hanger with a single line of fasteners parallel and on each side of the web. The fasteners are assumed to be stressed equally because of symmetry of the connection. An external tensile load on the connection will reduce the contact pressure between the T-stub flange and the base. However, depending on the flexural rigidity of the T-stub, additional forces may be developed near the flange tip. This phenomenon is

17.3 Bolt Groups Loaded in Tension—Prying Action

referred to as prying action and is illustrated in Fig. 17.4. Prying action increases the fastener force and may be detrimental to the strength and the performance of the fasteners.

If the flange of a T-stub connection is sufficiently stiff, the flexural deformations of the flange will be small compared to the elongation of the fasteners. Very little prying forces will be developed and the connection will behave much like a single bolt in tension. This is illustrated in Fig. 17.5a where the bolt force in a test specimen is plotted as a function of the external applied load. The maximum moment in the T-stub occurs at the interface between the web and the flange. Since very little prying force is developed, the flange is subjected to single curvature bending.

When more flexible T-stub flanges are used, the flexural deformation of the flange induces prying forces that result in additional bolt forces as illustrated in Fig. 17.5b. Initially the external load reduces the contact pressure between the flange and the base until separation at the bolt line occurs. Bending in the outer portions of the flanges develops prying forces acting between the bolt line and the edge of the flange as illustrated in Fig. 17.4. Yielding of the fasteners and the T-stub flange often permits an increase in the applied load with only a small increase in bolt force. Because of this plastic flow, the prying force is reduced at this load level (see Fig. 17.5b). Depending on the flexural rigidity of the flange and the properties of the fasteners, prying forces may persist up to the point of failure.

Test results have confirmed that both the stiffness properties of the flange and the fasteners are significant factors influencing the prying action.[16.2, 17.2-17.4] Other factors such as the magnitude of the initial clamping force of the fasteners, the grip length, and the number of lines of fasteners have also been studied. Test results have indicated that the initial

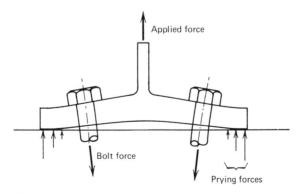

Fig. 17.4. Schematic of joint deformation.

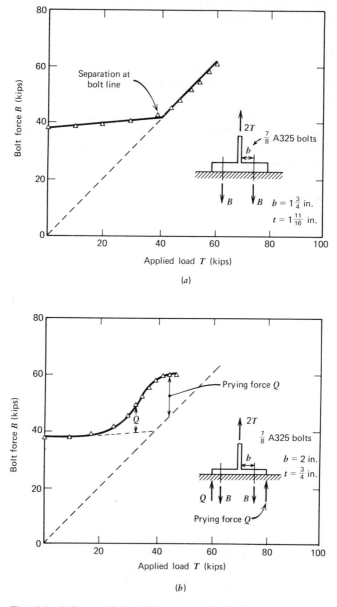

Fig. 17.5. Influence of plate thickness on applied load.

17.3 Bolt Groups Loaded in Tension—Prying Action

clamping force does not affect the prying action at ultimate load.[17.2, 17.3] This is illustrated in Fig. 17.6 where joints with two different bolt preloads are compared. The bolt force in the T-stub connection is plotted as a function of the applied load. The prying action at load levels close to the ultimate load were about the same for both conditions.

Although an increase in grip length may reduce the prying action at relatively low loads, the behavior at ultimate load is not significantly affected.[17.2, 17.3] The prying action at ultimate is influenced by the deformation capacity of the bolts. At ultimate load, the inelastic deformations of the threaded portion of the bolt are more critical than the small elastic elongations that occur in the bolt shank. An increase in grip length has only a minor effect as long as the length of the thread under the nut is relatively constant.

In the discussion so far it is assumed tacitly that the T-section is connected to a rigid base. However, practical situations do arise where the member to which the T-section is connected does not provide a rigid base. A typical example is a T-section which transfers the tensile component in a moment resistant beam-to-column connection. The web of the T-stub is connected to the beam tension flange and the flange of the T-section is bolted to the column flanges (see Fig. 17.1). If the column flanges do not

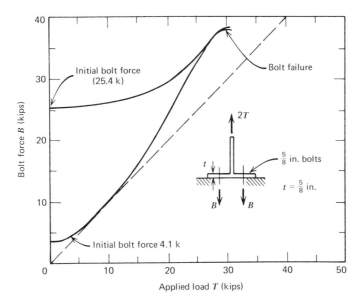

Fig. 17.6. Influence of initial bolt preload on prying action (Ref. 17.3).

provide adequate stiffness under the applied load system, the location of the prying forces may shift from the toe lines AB and CD, to the edges AC and BD (see Fig. 17.7a). In such connections the magnitude and the location of the prying forces are governed by the relative stiffnesses of the T-stub flange and the column flange. Generally, the resulting loading condition in such a connection is highly complex and has not been studied extensively. Reference 17.5 summarizes the results of a series of tests in which T-sections were bolted to the flanges of a wide flange shape. The T-sections were loaded in tension. The influence of the column flange thickness on the location and the magnitude of the prying forces was studied.[17.5] Some typical test results are shown in Fig. 17.7b. It is apparent from the deformation pattern that as the stiffness of the T-flange is increased, the prying forces tend to concentrate in the areas near the corners of the T-section. When the stiffness of the T-stub flange is much greater than the stiffness of the column flange, the T-section provides the rigid base and prying forces are developed because of deformations of the column flange.

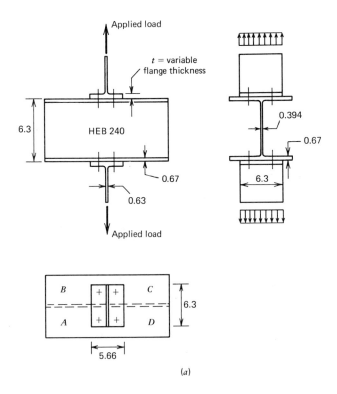

(a)

17.3 Bolt Groups Loaded in Tension—Prying Action

$t = 0.67$ in. $t = 0.79$ in.

$t = 0.98$ in. $t = 1.26$ in.

Fig. 17.7. T-stub sections bolted onto nonrigid support (Ref. 17.8). (a) Specimen dimensions. (b) Deformation pattern for various T-stub flange stiffnesses. (Courtesy of Stevin Laboratory Technical University Delft.)

When hangers have more than two rows of fasteners parallel to the web (see Fig. 17.8a), the effectiveness of the outer rows may be sharply reduced because of the flange flexibility. Tests have demonstrated that upon loading of the connection, the strain in the inner fasteners increases and continued to do so until failure occurs.[17.2] However, initially the strain in the outer bolts decreased slightly or remained constant. Thus in the early stages of loading, almost the entire load is carried by the inner bolts. Failure of the inner fasteners occurred before the strength of the outer fasteners could be developed. Increasing the flexural stiffness of the flange resulted in increased efficiencies. Test efficiencies between 45 and 80% were observed.[17.2] This shows that the outer bolts are not very effective in carrying the applied load unless the flanges are extremely heavy or stiffened as indicated in Fig. 17.8b.

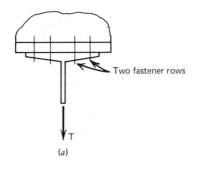 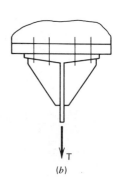

Fig. 17.8. Four-row hanger connections.

17.4 REPEATED LOADING OF TENSION-TYPE CONNECTIONS

As early as 1956 it was reported that prying forces could significantly reduce the fatigue strength of a tension-type T-connection.[17.1] Although extensive data are not available, further research has yielded information on the behavior of bolted T-connections under repeated loading conditions.[17.4] Fatigue tests were carried out on connections having a single line of fasteners on either side of the web. At the present time (1973), it is only possible to discuss qualitatively the behavior under repeated loading because a satisfactory theoretical solution is not available.

The bolt tension history of a single fastener installed in a plate assembly and subjected to an external tensile load was discussed earlier. The idealized relationship between the axial force in the bolt and the applied load is summarized in Fig. 17.3. The results plotted in Fig. 17.5a indicated that relatively stiff tension-type T-connections behave similarly to single bolt and plate assemblies. It is apparent that the increase in bolt force due to external applied load is small as long as the external load does not cause a separation of the plates. Hence providing an adequate initial clamping force reduces the stress range in the fastener under applied loads.

If the flanges of a T-connection loaded in tension are flexible, prying forces develop and a significant decrease in fatigue life results.[17.4] This decrease is dependent on the magnitude of the prying force. Typical data from tests with carbon steel (σ_y = 36 ksi) T-connections fastened by ¾-in. A490 bolts are summarized in Fig. 17.9. Similar connections fastened with A325 bolts exhibited the same behavior. It is apparent that an increase in prying force resulted in a decrease in fatigue life of the connection. These reductions can be qualitatively explained by examining the prying forces during a fatigue-type loading.

17.4 Repeated Loading of Tension-Type Connections

As illustrated in Fig. 17.5b, the prying force Q resulted in a large increase in bolt load as compared to the relatively rigid T-connection shown in Fig. 17.5a. The more flexible connection results in a greater stress range in the fastener. This decreases the fatigue strength of the connection. In addition, flexural deformations in the flange may distort the thread area of the bolt shaft. This also results in a higher stress range at the root compared to the average stress range in the bolt.

If the applied load on the connection is sufficient to produce yielding of the fasteners, a reduced clamping force results upon unloading. Subsequent cycles of load result in an increase in stress range. This is shown in Fig. 17.10 for a carbon steel T-connection fastened by $\frac{3}{4}$-in. A325 bolts.[17.4] An applied load of 24 kip/bolt, increased the bolt load by about 7 kip. Upon unloading, the initial clamping force was reduced from 32 to about 25 kip. When the external load was reapplied, the stress range during the second cycle was almost twice the stress range observed during the first cycle. A static test of an identical connection yielded a prying ratio Q/T equal to 0.37 at ultimate load.[17.4] When the same external load (24 kip/fastener) was applied to a connection in which very little prying force was developed,

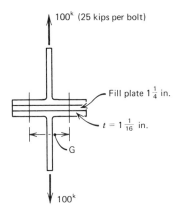

Fig. 17.9. Bolted T-stubs under repeated loading conditions (Ref. 17.4). Asterisk below denotes connection that did not fail. Test discontinued.

G (in.)	Applied Load Range per Bolt kips	Static Prying Ratio Q/T at Ultimate Load	Range in Average Bolt Stress First Cycle (ksi)	Number of Cycles to Failure
3	0–25	.02	2.2	3,000,000*
$4\frac{1}{2}$	0–25	.19	3.7	592,000
6	0–25	.45	10.4	32,000

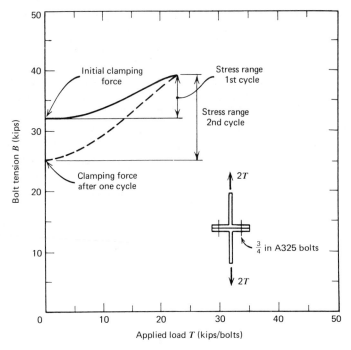

Fig. 17.10. Influence of prying force on fastener clamping force after unloading (Ref. 17.4).

the increase in bolt load was about 2 kip. The initial clamping force was not noticeably reduced after unloading. Subsequent cycles yielded a similar bolt load change. No marked difference was observed.

These studies illustrate that large prying forces decrease the static strength of the connection and also have a detrimental effect on the fatigue strength of the fasteners. It is apparent that a connection which develops little prying force is preferable under repeated loading.

17.5 ANALYSIS OF PRYING ACTION

Analytical and experimental studies of prying action have resulted in several mathematical models.[16.2, 17.3, 17.4, 17.6] Douty and McGuire used the model shown in Fig. 17.11 and suggested a formula based on an elastic analysis. They considered the properties of the bolts and the connected material and the geometry of the connection. These formulas were then modified to simplify application and reflect test results. The following semi-empirical equation was obtained.

$$Q = \left\{ \frac{\frac{1}{2} - (wt^4/30ab^2A_b)}{a/b[(a/3b) + 1] + (wt^4/6ab^2A_b)} \right\} T \qquad (17.5)$$

17.5 Analysis of Prying Action

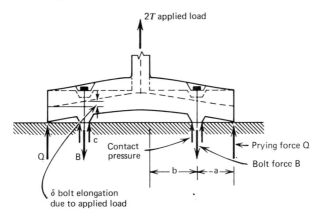

Fig. 17.11. Model used by Douty and McGuire.

This equation relates the prying force Q to the ultimate load of the connection. A similar formula with different coefficients was suggested for evaluating the prying force under working load conditions.[16.2]

Because of its complexity, Eq. 17.5 is not readily suited for design. The semi-empirical relationship for the prying force at ultimate was simplified in Ref. 17.6 to yield

$$\frac{Q}{T} = \left(\frac{3b}{8a} - \frac{t^3}{20}\right) \tag{17.6}$$

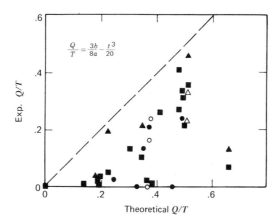

Fig. 17.12. Comparison between analytical and experimental results. ■ A325 bolts, $\sigma_{u\ spec}$ = 120 ksi. ▲ A490 bolts, $\sigma_{u\ spec}$ = 150–170 (ksi) ● 10k bolts, $\sigma_{u\ spec}$ = 142 ksi. ○ 4D bolts, $\sigma_{u\ spec}$ = 50 ksi. △ A502 rivets, σ_u 60–80 ksi.

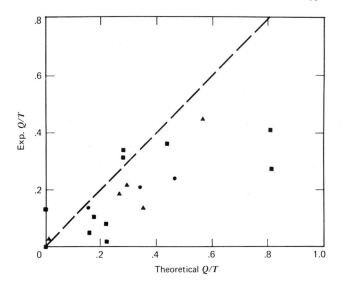

Fig. 17.13. Comparison between analytical and experimental results. ■ A325 bolts, ● 10k bolts: $Q/T = (100bd^2 - 18wt^2)/(70ad^2 + 21wt^2)$. ▲ A490 bolts, $Q/T = (100bd^2 - 14wt^2)/(62ad^2 + 21wt^2)$.

As is illustrated in Fig. 17.12, this equation tends to overestimate the prying force and provides conservative design results.[17.4]

An experimental and analytical study on connections consisting of two carbon steel T-stubs bolted together through the flanges with four A325 or A490 bolts was conducted at the University of Illinois, and resulted in empirical formulas to approximate prying.[17.4] The prying ratio Q/T at ultimate load for connections with A325 bolts was given as

$$\frac{Q}{T} = \left(\frac{100bd^2 - 18wt^2}{70ad^2 + 21wt^2}\right) \tag{17.7}$$

For connections with A490 bolts the coefficients 18 and 70 were replaced by 14 and 62, respectively. Test results were in slightly better agreement with the analytical results than provided by Eq. 17.6 as shown in Fig. 17.13. However, the empirical formulas are only applicable to the specific combination of bolt and plate material for which they were developed. Different formulas may be required for different bolt-plate material combinations.

A third analytical approach for predicting the prying force was suggested in Ref. 17.3. The simplified model, shown in Fig. 17.14, was used to

17.5 Analysis of Prying Action

describe the prying action in a T-stub with its flange bolted to a rigid base. The approach is not restricted to specific bolt-plate combinations since all major parameters which influence the prying action are included in the model. The Q denotes the prying force per bolt at ultimate and is assumed to act as a line load at the edge of the flange. Test results have shown this to be a reasonable assumption for conditions near ultimate as long as the edge distance a is within certain limits. The ultimate tensile load of the fastener is B, and the corresponding applied load per bolt is equal to T. The bending moment at the interface between the web and the flange is taken as M, and the moment at the bolt line due to the prying force Q is taken equal to $\alpha\delta M$ where δ is equal to the ratio of the net area (at the bolt line) and the gross area (at the web face) of the flange. The α represents the ratio between the moment per unit width at the centerline of the bolt line and the flange moment at the web face. When $\alpha = 0$, it corresponds to the case of single curvature bending, and $\alpha = 1$ corresponds to double curva-

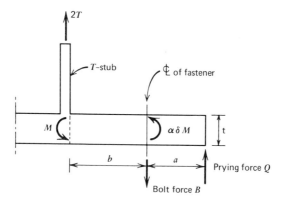

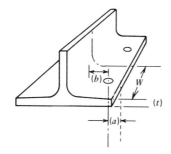

Fig. 17.14. Analytical model.

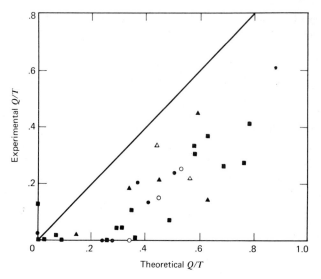

Fig. 17.15. Comparison between theoretical and experimental Q/T ratios. ■ A325 bolts, $\sigma_{u\,\text{spec}}$ = 120 ksi. ▲ A490 bolts, $\sigma_{u\,\text{spec}}$ 150–170 ksi. ● 1Ok bolts, $\sigma_{u\,\text{spec}}$ 142 ksi. ○ 4.D bolts, $\sigma_{u\,\text{spec}}$ 50 ksi. △ A502 rivets, σ_u 60–80 ksi.

ture bending. Note that the factor α is a function of the unknown ratio Q/T.

Moment equilibrium yields

$$(1 + \delta\alpha)M = Tb \tag{17.8}$$

where b is the distance from the centerline of the bolt to the web. The ultimate moment capacity of the gross area of the flange is

$$M = \tfrac{1}{4}wt^2\sigma_y \tag{17.9}$$

where σ_y is the yield point of the flange material, t the flange thickness, and w the length of the flange parallel to the web that is tributary to each bolt (see Fig. 17.14). Equilibrium at the bolt line yields

$$Qa = \alpha\delta\tfrac{1}{4}wt^2\sigma_y \tag{17.10}$$

Equilibrium of applied load, bolt force, and prying force is written as

$$B = T + Q \tag{17.10a}$$

17.5 Analysis of Prying Action

When expressed in terms of the other moment and equilibrium conditions this results in

$$B = T\left[1 + \frac{\delta\alpha}{(1+\delta\alpha)}\frac{b}{a}\right] \quad (17.11)$$

and

$$t = \left\{\frac{4Bab}{w\sigma_y[a + \alpha\delta(a+b)]}\right\}^{1/2} \quad (17.12)$$

Equation 17.12 relates the required flange thickness to the mechanical properties and geometrical dimensions of the constituent parts of the connection. Experimental results and the prying ratio Q/T obtained from Eq. 17.11 are compared in Fig. 17.15 for different types of bolts. A few data, obtained from riveted specimens, are included as well.

It is apparent that the solution given by Eqs. 17.11 and 17.12 overestimates the prying force. The variation is comparable to Eqs. 17.6 and 17.7. Among the factors causing the difference between the load transfer predicted by the idealized model and the test results are strain hardening and the actual distribution of forces. The model assumes the bolt force B to act at the center-line of the bolt. As a result of flexural deformations in the flange, the bolt force B is acting somewhere between the bolt axis and the edge of the bolt head, as indicated in Fig. 17.16. This decreases the distance b and changes the prying ratio Q/T directly. To approximate this assumption, a revised equilibrium condition was developed using modified distances a' and b' defined in Fig. 17.17b. The predicted prying force based

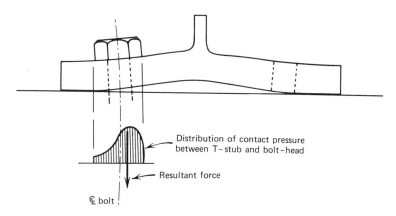

Fig. 17.16. Influence of flange deformations on location of resultant bolt force.

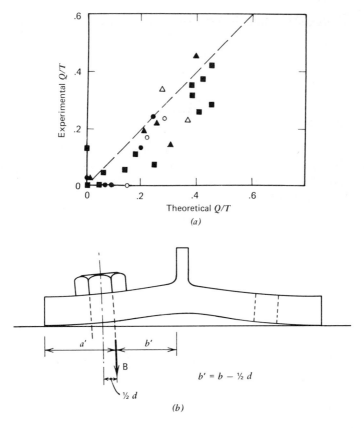

Fig. 17.17. Comparison between analytical and experimental Q/T ratios for modified a and b distances. ■ A325 bolts, $\sigma_{u\ spec}$ 120 ksi. ▲ A490 bolts, $\sigma_{u\ spec}$ 150–170 ksi. ● 10k bolts, $\sigma_{u\ spec}$ 142 ksi. ○ 4D bolts, $\sigma_{u\ spec}$ 50 ksi. △ A502 rivets, σ_u 60–80 ksi.

on these modified dimensions provides better agreement with the test results as illustrated in Fig. 17.17a. As noted in Fig. 17.17a, using these modified dimensions is likely to result in a conservative design of the bolts, since the model tends to overestimate the influence of the prying force. The resultant fastener force B was assumed to act at a distance b' equal to $b - d/2$ from the web face. The distance a' was taken equal to $a + d/2$. The model assumes the prying force Q at ultimate load to be a line load at the tip of the flange.

Tests have indicated that this is a reasonable assumption as long as the end distance is not much greater than the distance b. Therefore, the end distance a should be limited to $1.25b$.

17.6 DESIGN RECOMMENDATIONS

17.6.1 Static Loading

Several semi-analytical and empirical approximations for the prying force in T-connections with a single line of fasteners on each side of the web have been examined. All of the methods provided about the same degree of fit to the test data.

A modification of the equilibrium method proposed by Struik and deBack[17.3] was observed to have several advantages. Of primary importance was the fact that it was applicable to a wide range of fasteners and steel and readily suited for design. The analytical model used by Douty and McGuire had several coefficients adjusted on the basis of experimental work.[16.2] Hence it was not directly applicable to a variety of fasteners and materials. The empirical formulas developed by Nair *et al.* were only applicable to specific plate and bolt combinations.

Although only a few tests on connections with carbon steel bolts or rivets were available, the experimental data are in reasonable agreement with Eqs. 17.11 and 17.12 as illustrated in Fig. 17.17. The simplified model provides a satisfactory basis for designing bolted and riveted tension-type T-connections.

Connections with more than two gage lines of fasteners are not effective unless special provisions such as additional stiffening of the flange is provided.[17.2] If this is not provided, the load capacity is provided largely by the inner fasteners alone.

i. Allowable Stress Design. The minimum tensile capacity of a fastener is equal to the product of the fastener stress area A_s and its minimum specified tensile strength $\sigma_{u\,\text{spec}}$ in kilopounds per square inch. As noted in Chapter 4 the tensile capacity of a bolt can be expressed in terms of the nominal bolt area A_b as

$$B_{u\,\text{spec}} = 0.75\,A_b\sigma_{u\,\text{spec}} \qquad (17.13)$$

Applying a factor of safety with respect to ultimate load equal to 2.0 yields an allowable tensile load B_{all}, per fastener as

$$B_{\text{all}} = (0.5)(0.75)A_b\sigma_{u\,\text{spec}} \qquad (17.14)$$

or

$$B_{\text{all}} = 0.375 A_b\sigma_{u\,\text{spec}} \qquad (17.15)$$

A factor of safety of 2.0 is consistent with previously used values. It is also compatible with allowable shear and bearing stresses for bolts.

To provide a uniform margin between working load and ultimate strength, the applied load and prying force should not exceed the allowable bolt load. Hence

$$B_{\text{all}} \geq T + Q \qquad (17.16)$$

The prying force Q depends on the geometrical dimensions of the connection as well as upon the applied load T. These factors determine the value of α which is in turn related to the prying force Q as given in Eq. 17.10. The design recommendations, summarized hereafter, can be used either for analysis or for design purposes. For design, tentative dimensions of the T-stub must be selected together with a bolt size capable of carrying the applied load T. On the basis of these initial components, the maximum acceptable α can be determined from Eq. 17.17 (see design recommendations). The maximum acceptable value of α for design is 1.0 as this is equivalent to double curvature bending with plastic hinges forming at the bolt line and the interface between the web and the flange. If the maximum value of α determined from Eq. 17.17 exceeds 1.0, a value of $\alpha = 1.0$ is selected as this constitutes the limiting state. Such a situation indicates that even when two plastic hinges have formed on either side of the flange, the fasteners are still not loaded to their full tensile capacity. In other words, the fasteners are overdesigned. To reflect this overcapacity of the fasteners, the value of $\bar{B}$ in Eq. 17.18 should be equal to the maximum bolt force at ultimate multiplied by a factor a safety of 2.0. The bolt load at failure can be directly obtained from Eq. 17.17 for the condition $\alpha = 1.0$. The required flange thickness is then determined on the basis of this fastener load and not the tensile capacity of the fastener. The fasteners tensile capacity cannot be developed because of limiting flange capacity. Once the value of α is determined, it can in turn be substituted into Eq. 17.18 together with the various components, and the minimum flange thickness required for the selected bolt size and geometrical dimensions can be determined. The suitability of the assumed T-stub can then be evaluated.

Equations 17.17 and 17.18 are also applicable for analysis purposes. Equation 17.18 can be used to evaluate the coefficient α. Upon substitution of α into Eq. 17.17, the total bolt force can be evaluated. If the bolts are inadequate, either the geometrical dimensions or the bolt size can be altered.

DESIGN RECOMMENDATIONS FOR T-CONNECTIONS UNDER STATIC LOADING CONDITIONS

Allowable Stress Design

Allowable tensile load per fastener

$$B_{\text{all}} = 0.375 A_b \sigma_{u \text{ spec}}$$

17.6 Design Recommendations

Check adequacy of fastener to resist the applied load and prying action:

$$B_{\text{all}} \geq T + Q$$

or upon substituting Eq. 17.11 with modified a and b distances:

$$B_{\text{all}} \geq T\left[1 + \frac{\delta\alpha}{(1+\delta\alpha)}\frac{b'}{a'}\right] \qquad (17.17)$$

The T-flange thickness must be equal to or exceed:

$$t = \left\{\frac{4\bar{B}a'b'}{w\sigma_y[a' + \delta\alpha(a'+b')]}\right\}^{1/2} \qquad (17.18)$$

where $a' = a + d/2$

$b' = b - d/2$

$\bar{B}$ = estimated fastener load at failure of the connection

if $\quad \alpha < 1.0; \quad \bar{B} = 0.75 A_b \sigma_u$

if $\quad \alpha \geq 1.0$ it is taken as 1.0 and

$$\bar{B} = 2T\left[1 + \frac{\delta}{(1+\delta)}\frac{b'}{a'}\right]$$

Maximum value of distance a

$$a \leq 1.25b$$

The design recommendations given in this section are valid for tension-type connections fastened to a rigid base. It was noted in Section 17.3 that the stiffness of the base to which the T-section is connected is an important parameter in the development of prying forces. If the base does not provide enough stiffness, the fastener loads and prying forces should be evaluated on the basis of the geometrical dimensions and material properties of the flange to which the T is connected. The joint component which provides the least stiffness results in the greatest prying forces and governs the design of the fasteners.

ii. Load Factor Design. The design of T-connections by load factor design is directly comparable to allowable stress design. The only difference is that the load on the fastener at the factored load level should not exceed the ultimate tensile load of the fastener multiplied by a reduction factor Φ. A reduction factor Φ equal to 0.85 is in reasonable agreement with past practice. A load factor of 1.7 and a reduction factor of 0.85 yields a design which is comparable to allowable stress design.

DESIGN RECOMMENDATIONS FOR T-CONNECTIONS UNDER STATIC LOADING CONDITIONS

Load Factor Design

$$B = 0.75 A_b \sigma_u$$

Maximum tensile capacity of fastener
Check adequacy of fastener to resist the applied load as well as prying action

$$T'\left[1 + \frac{\delta\alpha}{(1+\delta\alpha)}\frac{b'}{a'}\right] \leq \Phi B$$

where reduction factor $\Phi = 0.85$.
T' represents the applied load per fastener at the factored level.
The T-flange thickness is given by

$$t = \left\{\frac{4\bar{B}a'b'}{w\sigma_y[a' + \delta\alpha(a'+b')]}\right\}^{1/2}$$

where $a' = a + d/2$
$b' = b - d/2$
$\bar{B}$ = the fastener load at the factored load level

if $\alpha < 1.0$

$$\bar{B} = B = 0.75 A_b \sigma_u$$

if $\alpha \geq 1.0$ it is taken as 1.0 and

$$\bar{B} = T'\left[1 + \frac{\delta}{(1+\delta)}\frac{b'}{a'}\right]$$

Maximum value of distance a

$$a \leq 1.25b$$

17.6.2 Repeated Loading.

The fatigue strength of T-connections in tension is significantly affected by the clamping force of the fastener. Therefore, in situations where repeated loading is expected, special attention must be directed to bolt installation procedures to ensure that the bolts are properly tightened and provide the desired clamping force.

As noted earlier, prying forces in T-connections lead to severe reductions in fatigue strengths. To avoid a reduction in strength and substantial decreases in life, the T-connection should be dimensioned so that prying forces are minimized. This can be accomplished by providing a reasonable rigid T-connection as was shown in Fig. 17.5a. This will ensure that the applied load does not cause separation of the plates. Consequently, the bolt forces will only experience a small change in stress.

The design of the connection should not permit prying forces to be developed. This can be achieved by setting the factor α equal to zero.

DESIGN RECOMMENDATIONS FOR THE T-CONNECTIONS REPEATED TYPE LOADING

Allowable tensile load on fastener:

$$B_{\text{all}} = 0.375 A_b \sigma_{u\text{ spec}}$$

Check adequacy of bolts:

$$T \leqq B_{\text{all}}$$

Provide flange thickness which does not induce prying action:

$$t = \left\{ \frac{4Bb'}{w\sigma_y} \right\}^{1/2}$$

where $b' = b - d/2$
and $B = 0.75 A_b \sigma_u$

References

17.1 W. H. Munse, "Research on Bolted Connections," *Transactions, ASCE*, Vol. 121, 1956, p. 1255.

17.2 W. H. Munse, K. S. Peterson, and E. Chesson, Jr., "Strength of Rivets and Bolts in Tension," *Journal of the Structural Division, ASCE*, Vol. 85, No. ST3, March 1959.

17.3 J. H. A. Struik and J de Back, *Tests on Bolted T-Stubs with Respect to a Bolted Beam-to-Column Connections,* Report 6-69-13, Stevin Laboratory, Delft University of Technology, Delft, the Netherlands, 1969.

17.4 R. S. Nair, P. C. Birkemoe, and W. H. Munse, *High Strength Bolts Subjected to Tension and Prying*, Structural Research Series 353, Department of Civil Engineering, University of Illinois, Urbana, September 1969.

17.5 J. de Back and P. Zoetemeyer, *High Strength Bolted Beam-to-Column Connections, The Computation of Bolts, T-Stub Flanges and Column Flanges*, Report 6-72-13, Stevin Laboratory, Delft University of Technology, Delft, The Netherlands, 1972.

17.6 ASCE, *Commentary on Plastic Design*, Manual 41, New York, 1971.

17.7 J. L. Rumpf, "Riveted and Bolted Connections," in *Structural Steel Design*, Ronald Press, 1964, Chap. 18.

Chapter Eighteen

Beam-To-Column Connections

18.1 INTRODUCTION

Beam-to-column connections play an important role in the load partition of structural frames. The major function of these connections is to transfer the loads that are applied to the beams and the floor system to the columns. In its simplest form the connection is only used to transfer the end reaction of the beam to the column, and the beam is assumed to be simply supported. If restraints are provided, the end rotations of the beam are minimized, and the maximum positive moment in the beam can be reduced by the resulting end moments. Connections of this nature are often referred to as moment-resistant joints. Connections that are only capable of transferring the reaction of the beam are called shear connections.[18.1]

The behavior of beam-to-column connections is of major interest to engineers and a significant amount of research has been done or is underway. These studies are aimed at developing and improving design rules for beam-to-column connections.[16.1-16.3, 18.1-18.7] These projects all focus on the general requirements for connections which can be summarized as: (1) sufficient strength, (2) adequate rotation capacity, (3) sufficient stiffness, and (4) economic fabrication.

Most of the past research on beam-to-column connections was performed on welded or riveted specimens. However, as the advantages of bolted connections and combination bolted and welded connections became more apparent because of decreased fabrication and erection costs, research on these types of connections was increased.[14.4, 18.7]

Shop connections are often welded and field connections are bolted in current practice. As a result of these fabrication procedures, a wide variety of beam-to-column connections are encountered in the field.[13.10, 15.6] It is still not possible to describe and predict accurately the behavior of many of these connections because of their complexity. This chapter summarizes the present state of knowledge and provides guidelines for design. The design recommendations for these joints are based on available information and result in a conservative, safe design. Additional experimental and theoretical work is needed before more liberal and improved design rules can be developed.

18.2 CLASSIFICATION OF BEAM-TO-COLUMN CONNECTIONS

Depending on their rotational characteristics, beam-to-column connections are classified as flexible, semi-rigid, or rigid connections.[18.1] Flexible connections are also called shear connections, and the semi-rigid and rigid-type connections are often referred to as moment-resistant connections.

The rotational characteristics of beam-to-column connections are important to the engineer as they affect the required beam size. For idealized rigid joints, the beam size is generally governed by the fixed end moment, $M = wl^2/12$, for a uniformly loaded beam. If the same beam is attached to the column by a flexible-type shear connection, the maximum moment for a simply supported beam becomes $M = wl^2/8$. Actual situations in the field will generally be somewhat less rigid than assumed for the rigid connection, and somewhat more rigid than assumed for the flexible connection. The classification of a connection depends entirely on the joint geometry and loading conditions. Generally, it is not possible to define how a joint should be classified unless test results and experience are available.

The simplest type of beam-to-column connection is the flexible connection which provides relatively low resistance against rotations. Hence the connection mainly transfers shear to the column. Typical examples that fall into this category are the web angle (or standard beam) connection, web structural tee, and seat angle connections, shown in Fig. 18.1a. The structural T-connections, end plate connections, and flange plate connections, shown in Fig. 18.1c, are typical examples of high-moment-resistant beam-to-column connections. By combining web angles or a T-section with a beam seat and tension flange plate or angle, a semi-rigid connection results with a greater moment resistance than the flexible connection. Unfortunately the degree of restraint is often difficult to evaluate unless test data are available.

Typical moment rotation characteristics for several types of beam-to-column connections are shown in Fig. 18.2. These relationships combined with the beam line concept (introduced in Ref. 18.1) are often used to estimate the moment that will be developed by a particular connection, span, and beam size. The beam line defines the relationship between the end moment and end rotation of a beam. If a beam is uniformly loaded and subjected to restraining end moments, M, the end slope ϕ is equal to

$$\phi = \frac{1}{24}\left(\frac{wl^3}{EI}\right) - \frac{Ml}{2EI}$$

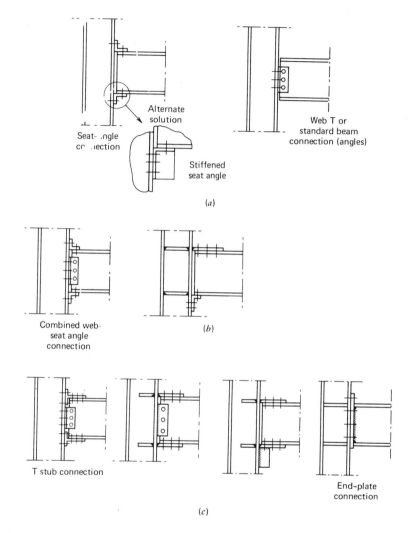

Fig. 18.1. Types of beam-to-column connections. *Note.* The need for column stiffeners in any of these connections must be checked. (*a*) Flexible connections; (*b*) semi-rigid connections; (*c*) rigid connections.

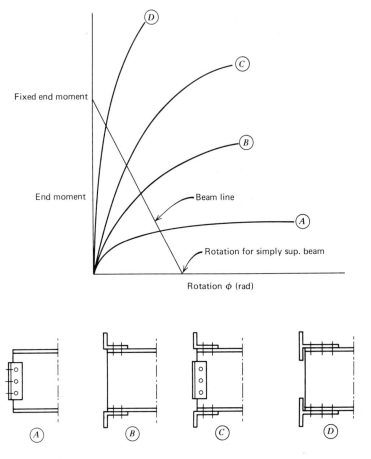

Fig. 18.2. Typical moment rotation curves and beam-lines (Ref. 13.10).

This relationship is plotted in Fig. 18.2. The intersection of the beam line and moment rotation curves for the various connections indicates the moment resistance expected under these conditions. For example, the standard web angle connection (connection A in Fig. 18.2) develops about 20% of the fixed end moment for this particular combination of beam and connection geometry. The same connection with added top and seat angles (connection C) develops about 75% of the fixed end moment.[18.1]

18.3 BEHAVIOR OF BEAM-TO-COLUMN CONNECTIONS

The stiffness and strength of beam-to-column connections are closely interrelated and of major importance to the performance of the connection.

18.3 Behavior of Beam-to-Column Connections

Strength requirements ensure that the connection has the ability to transfer the anticipated loads. Stiffness requirements relate to the ability to develop the desired restraint or lack of restraint. To meet the stiffness and strength requirements, additional stiffening of the column web or flanges may be needed, since certain joint components are subjected to highly localized concentrated forces. Stiffeners are often necessary to prevent crippling of the column web in the compression region, excessive yielding of the column web, or deformation of the column flange near the tension flange of the beam. If the shear capacity of the column web is critical, shear stiffening may be required for that purpose as well.

The load-deformation characteristics and approximate methods of analysis for typical beam-to-column connections are discussed in this section. Features from different types of connections are sometimes combined to meet the design requirements. Only the strength aspects of the connection are discussed in this section. Problems related to stiffening of the column web are treated separately in Section 18.4.

18.3.1 Flexible Beam-to-Column Connections

The web angle or standard beam connection, as well as the seat angle connection, are typical flexible beam-to-column connections. Generally they are assumed to be completely flexible and capable of transferring shear alone. To justify these assumptions, the connections must allow for ample end rotation.

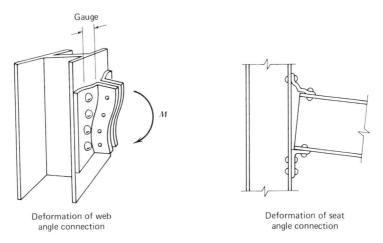

Fig. 18.3. Deformations of flexible beam-to-column connections.

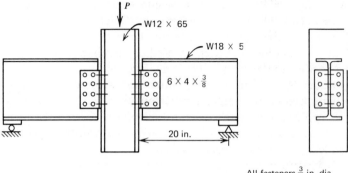

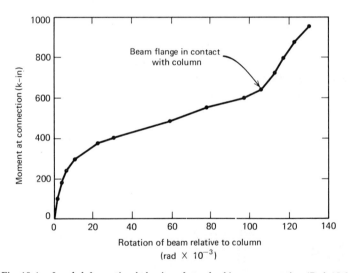

Fig. 18.4. Load-deformation behavior of standard beam connection (Ref. 18.2).

The rotation capacity of the connection is largely governed by the deformation capacity of the angles, as shown in Fig. 18.3. Experiments have indicated that most of the rotation of the connection comes from the deformation of the angles; fastener deformations play only a minor role.[18.1, 18.2] To minimize rotational resistance, the thickness of the angle should be kept to a minimum and a relatively large gage, g, provided (see Fig. 18.3).

18.3 Behavior of Beam-to-Column Connections

A typical moment-rotation diagram for a bolted and riveted standard web connection is shown in Fig. 18.4. In this test, the heels of the angles on the tension side began to separate from the column flanges at about 260 kip-in. The toes of the angles remained in contact with the column. Yielding of the angles decreased the rotational resistance. After the compression beam flanges had made contact with the column flanges, the moment resistance of the connection increased as shown in Fig. 18.4. Failure of the connection occurred from excessive yielding and tearing of the connection angles (see Fig. 18.5).

From these test series it was concluded that web angle beam-to-column connections offer some resistance to rotations at the ends of the beam. This partial restraint is relatively small and estimated to be about 10% of the fixed end moment provided by rigid moment-resistant connections.[18.2, 18.9] Rotation restraints of the same order of magnitude can be expected in seat angle connections as well.[18.3]

Most web angle connections are checked for their shear carrying capacity alone. This is governed by either the shear capacity of the fasteners, the shear capacity of the angles, or the bearing capacity of the beam web, column flanges, or angles. In evaluating the shear capacity of the angles, the influence of the end moment is neglected. Also, fasteners are assumed

Fig. 18.5. Angle failure in standard beam connection described in Fig. 18.4. (Courtesy of University of Illinois.)

to be subjected to shear alone. Tensile stresses introduced by the deformed angle and the end moment (see Fig. 18.3) are neglected.

The upper angle in a seat connection (see Fig. 18.3) is mainly used to provide lateral stability for the beam. This joint component is not considered as load carrying. The total shear force is assumed to be transmitted to the column by shear on the fasteners in the seat angle. The thickness of the seat angle is governed by critical bending stress on the outstanding leg. The usual practice is to consider the stress at the toe of the fillet of the outstanding leg. The required angle thickness is determined from the bending moment at that section. The reaction is assumed to act at midpoint of the bearing length.[13.10]

18.3.2 Semi-Rigid Connections

A combination web angle and seat angle connection results in significant increases in the joint restraint characteristics. Depending on the dimensions of the joint components and the loading conditions, these combination joints are sufficiently stiff to result in a substantial reduction in the midspan moment of a beam.[18.1] Beam-to-column connections of this type (see Fig. 18.1b) are classified as semi-rigid.

Little experimental evidence is available on the load-deformation behavior and load partition for this type of connection.[18.1] Since the behavior of the connections is complex and because of the lack of experimental data, a simplified conservative approach is used for design. Current practice assumes that the web angles will carry the shear. Thick top and bottom angles are used to transfer the end moment of the beam. Connections designed on the basis of these assumptions have provided satisfactory performance.

The shear connection design is identical to the web angle connection discussed in Section 18.3.1. Both angles in the semi-rigid connection are

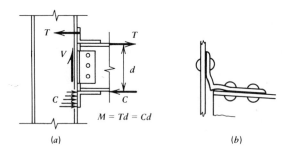

Fig. 18.6. Assumed behavior of semi-rigid connection.

18.3 Behavior of Beam-to-Column Connections

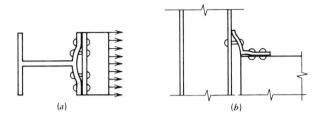

Fig. 18.7. Influence of deformations on fastener elongations.

considered to be load-carrying components which was not the case for seat angle connections. Both angles are subjected to bending forces. However, the angle which connects the beam tension flange to the column flange is the critical one. A typical deformation condition for the tension angle is shown in Fig. 18.6b. Depending on the stiffness of the angle, prying forces may develop near the toe of the outstanding leg. Therefore, it is desirable to consider the influence of prying forces on the bending stress in the angle and the fastener tension. For analysis, the angle can be assumed to act like a T-stub connected to a rigid base and loaded in tension. This provides a conservative design as it assumes the angle to be fastened to a rigid base. Since the angle is fastened to a column flange, the decreased stiffness tends to relieve part of the restraint supplied by the angle. In general, the forces developed in a semi-rigid connection cannot be reasonably approximated unless a test is conducted. This permits the stiffness and distribution of the forces in the connection to be evaluated.

The moment capacity of the connection is limited by the number of fasteners which can be placed in a single transverse line in the vertical leg of the angle connecting the tension flange to the column flanges. Because of deformation of the column flange (see Fig. 18.7) only the first fasteners on each side of the beam web may be fully effective in transferring the forces. Stiffening of the column flanges may be required unless they are at least as thick as the angle.

18.3.3 Rigid Connections

Replacing the angles of a combined web-seat angle connection (see Fig. 18.1b) with structural T-sections results in a connection with significantly improved moment rotation characteristics. Such a connection (see Fig. 18.1c) provides a rigid joint with high rotational resistance. The increase in rotational resistance is provided by the symmetrically loaded T-sections. Unlike angle connections, which are connected to the column flanges by two or more fasteners on one line, the T-section allows two or more fasten-

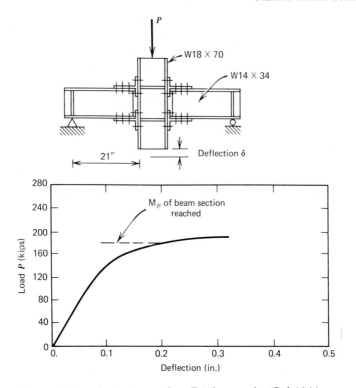

Fig. 18.8. Load-deflection curve for a T-stub connection (Ref. 16.1.)

ers to be used effectively on two lines to transfer the tensile forces that result from the applied moment. This results in an increase in moment capacity and joint stiffness. Since the T-sections are symmetrically loaded, they do not permit as much deformation to occur as compared to eccentrically loaded angles (see Fig. 18.3).

The design of the T-stub connection utilizes assumptions similar to those used for combined web-seat angle connections. The flange connection is assumed to transfer the moment and the shear force is transferred by the web connection. Tests were carried out on connections of this type to evaluate the validity of these assumptions.[16.1, 16.2] Typical test results are illustrated in Fig. 18.8. The effect of beam shear and the presence of the web angles on the behavior of the flange connections was investigated. In addition, these tests yielded valuable information on the rotation capacity of these connections.

18.3 Behavior of Beam-to-Column Connections

The test results indicated that the behavior of the bolts connecting the T-stubs to the beam flanges was similar to the behavior observed with simulated flange plate splice tests.[16.2] The connections exceeded the plastic moment of the gross cross-sectional area of the beam, despite the presence of the holes in the flanges. Substantial rotational capacity was attained (see Fig. 18.8) when premature failure of the joint components was prevented. It was further concluded that the beam shear had no significant effect on the performance of the connection. The shear was largely carried by friction between the T-stubs and the column flanges. There was very little difference in bolt tension in the bolts connecting the tension T-stub, regardless of the magnitude of the prying forces.[16.2]

The test results generally supported the assumptions made in design. Although some shear can be transferred by the web of the T-stub, web angles are needed to assist with the shear transfer. This is particularly true if large shear forces exist.

End plates welded to the beam cross-section have been used in beam-to-column connections and butt-type beam splices (see Chapter 16). Two types of end plates are used as shown in Fig. 18.9. In one type the fasteners are placed between the beam flanges, and in the other type the end plate is

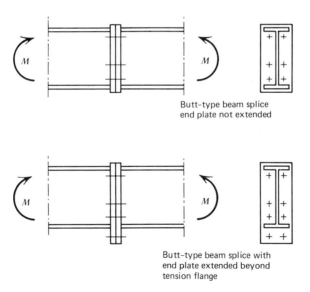

Fig. 18.9. End plate types. *Note.* Connect end plates to beams with enough weld to develop full bending strength of beam.

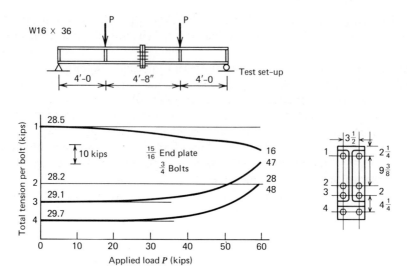

Fig. 18.10. Bolt force versus applied load (Ref. 16.2.)

extended beyond the tension flange and fasteners are centered around the flange.

The exact load transfer in this type of connection is complex. The shear forces acting on the connection are transferred by frictional resistance and/or by shear on the fasteners. The fasteners are also subjected to tensile loads which resist the bending moment. The forces in the bolts change under the applied loads and are dependent on the magnitude of the initial bolt tension.

Several experimental studies were made to examine the load-deformation behavior of this type of connection and to develop design rules.[16.1, 16.2, 18.4-18.6] These studies have indicated that, by proper dimensioning of the joint components, this type connection can transmit the shear and bending forces. Friction developed between the end plate, and the element to which it is connected can resist the shear. The bolts mainly effective in resisting the tension flange force are those adjacent to the tension flange. This is illustrated in Fig. 18.10 where the bolt forces in a moment splice end plate connection are plotted as a function of the applied load.[16.2] It is apparent that the fasteners adjacent to the tension flange were most effective and almost equally loaded. There was no appreciable change in tension in the fasteners located in row 2 at any load level. It was concluded from Fig. 18.10 that the variation of the force in the several rows of a bolt pattern depends primarily on the stiffness of the end plate and

18.3 Behavior of Beam-to-Column Connections

whether the plate yields before fracture of the critical fasteners takes place. At first, strains will increase in proportion to the distance of the fasteners from the compression flange. Because of the strain gradient, differences in bolt loads result but decrease as plastic deformations of the bolt develop. If the bolts have sufficient ductility, all bolts in the tension region will develop the same capacity at ultimate load.[16.2] Unless it is sufficiently thick, the end plate will yield and a linear strain distribution does not occur. This is apparent in Fig. 18.11 which shows an end-plate connection after failure.[18.5] The pressure distribution at the interface of the end plate and the column is shown in Fig. 18.12 and indicates that prying forces were developed at the edges of the end plate near the tension flange.[18.5]

Fig. 18.11. End plate connection after failure. (Courtesy of University of Sheffield).

Test results have shown that the bolts which are effective in resisting the moment for flexible end-plate connections are adjacent to the tension flange. The connection is flexible if prying forces are developed at the edge of the end plate in the tension region. If a connection is designed such that no prying forces are developed, a linear strain distribution among the fastener rows can be assumed and the inner fasteners may contribute to the capacity of the connection. The ultimate moment resistance of the connection is the summation of the products of the effective fastener loads and their respective distance from the center of rotation. At the ultimate load the center of rotation is near the center line of the compression flange. This is compatible with existing experimental observations.[16.2, 18.4, 18.5]

The bolts and end plate adjacent to the tension flange can be conservatively designed by assuming that they are equivalent to a T-stub connection loaded in tension. Design procedures for this idealization are given in Chapter 17.

Although the primary transfer of shear is concentrated near the compression side of the joint, it can be conservatively assumed that all fasteners carry an equal part of the shear load. Hence the fasteners in an end-plate connection are subjected to combined shear and tension.

The magnitude of initial clamping force does not influence the ultimate strength of the connection. It does influence the shear resistance of slip-resistant joints.

When end plates do not extend beyond the tension flange, their behavior is not well known because available data are not extensive. In general, these types of end-plate connections are less efficient and require thicker end plates. Reference 16.2 suggested that end plates that do not extend beyond the tension flange should be proportioned to resist a moment equal to the product of the beam flange force and the distance between the center of the beam flange and the nearest row of bolts. Plate thicknesses determined in this manner appear to provide a linear variation in fastener strain throughout the connection depth. Additional test data are needed to verify this suggested method for a range of sizes.

18.4 STIFFENER REQUIREMENTS FOR BOLTED BEAM-TO-COLUMN CONNECTIONS

The full capacity of a moment resisting beam-to-column connection can only be developed if the column does not exhibit premature failure. The column is subjected to highly localized forces from the applied moments and can deform as shown schematically in Fig. 18.13a. Excessive deformations of connected parts should be avoided. There are two major effects of the beam flange forces which have to be examined because they may result

18.4 Stiffener Requirements for Bolted Beam-to-Column Connections

in excessive deformations. On the compression side of the beam, crippling of the column web can occur. On the tension side, excessive yielding and distortion may result in fracture of the column web or bolts. Web buckling is illustrated in Fig. 18.13b where an end-plate connection at ultimate load is shown. Because of the lack of stiffening in the compression region, the column web buckled and the connection could not develop the plastic moment capacity of the beam.[18.4]

Several investigators[18.4-18.6] have examined the stiffening requirements for bolted beam-to-column connections. Since many joint geometries and boundary conditions exist, the problem is extremely complex and no satisfactory design approach is available at the present time (1973). Often the requirements developed for stiffening welded beam-to-column connections are used.[18.8] Since the concentrated forces are more localized in welded connections, their application to bolted connections results in a conservative design. Pending further research, criteria based in part on the requirements used for welded beam-to-column connections are reasonable.

The requirements for stiffening of the column are summarized as follows. As proposed in Ref. 18.8, the compression flange force on the column

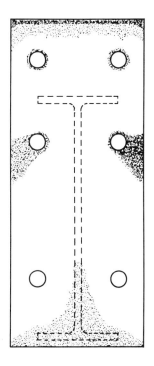

Fig. 18.12. Pressure distribution at interface as recorded on interposed paper backed up by carbon paper (Ref. 18.5.)

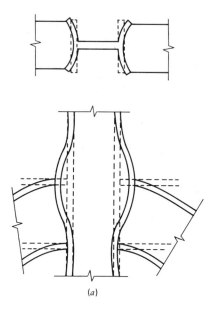

Fig. 18.13. Deformation of column in moment resistant connection. (a) Distortion of unstiffened column. (b) Web crippling in beam-column connection. (Courtesy of British Steel Corp.)

18.4 Stiffener Requirements for Bolted Beam-to-Column Connections

is assumed to be distributed on a 2.5:1 slope from the point of contact to the column k line (see Fig. 18.14). If the compression flange force is distributed to the column flange by either an end plate or a structural T-section, it can be assumed to be distributed over a region on the column face about twice as great as the beam flange thickness. Hence the force in the beam flange is assumed to be resisted by a length of column web equal to $(Q + 5k_c)$, where Q is the sum of the beam flange thickness and twice the end-plate thickness (for the plate connection) or the web thickness of the T-stub and twice its flange thickness, and k_c the column fillet depth. For equilibrium, the resistance of the effective area of the web must equal or exceed the applied concentrated force of the beam tension or compression flange. This yields the following condition

$$\sigma_{yc} w_c (Q + 5k_c) \geq A_f \sigma_{yb} \tag{18.1}$$

where w_c is the thickness of the column web and A_f the flange area of the beam. The yield point of the column web is given by σ_{yc} and the yield point of the beam flange by σ_{yb}. If the column web resistance is less than provided by Eq. 18.1, stiffeners are required.

If flange splice plates are welded to the column on the compression or tension side of the beam, the provisions developed for welded connections are directly applicable.[18.8] The force from the compression flange is resisted by a length of the column web equal to $(t_s + 5k_c)$ where t_s is the splice

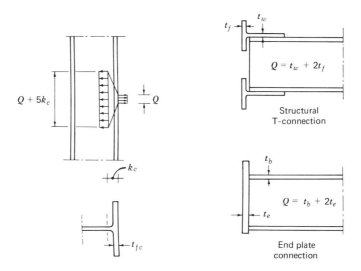

Fig. 18.14. Assumed distribution of compression flange force in bolted beam-to-column connections.

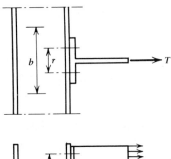

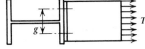

Effective length b: $b = r + \frac{3g}{2}$

Fig. 18.15. Effective length of column flange.

plate thickness. For the tension flange, Ref. 18.8 has shown that the column flange provides adequate resistance against excessive deformations from the concentrated forces delivered by the tension splice plate if

$$t_{fc} \geq 0.4 \left(A_f \frac{\sigma_{yb}}{\sigma_{yc}} \right)^{1/2} \tag{18.2}$$

where t_{fc} is the column flange thickness. Tests of welded connections proportioned to these recommendations indicated that the connections were able to develop the full plastic moment of the beam.[18.8]

If a T-section or an end plate is bolted to the column flange, the concentrated tension force is distributed into the column flange by the fasteners. The system of applied forces differs significantly from the case of the splice plate welded to the column. The application of Eq. 18.2 is likely to yield over conservative results. Current European specifications incorporate recommendations regarding stiffener requirements in these situations.[5.30] The need for additional column stiffening is estimated on the basis of an idealized model of the column flange behavior. The column flanges can be considered satisfactory if the moment acting on an effective length b is within certain limits. The idealized force system is shown in Fig. 18.15. Equilibrium on the assumed critical flange section yields

$$\left(\frac{T}{2}\right)\left(\frac{g}{2}\right) \leq \bar{M} \tag{18.3}$$

where T is the applied force and g the fastener gage. $\bar{M}$ is the permissible

18.5 Design Recommendations

moment on the effective length of column flange b. The effective length b is defined as

$$b = r + \frac{3g}{2} \tag{18.4}$$

where r is the fastener pitch or spacing along the flange.

Column stiffeners should be proportioned to carry the excess concentrated flange forces that the column web and flange are unable to resist.

If a single beam frames into a column or if the moment from two beams at an interior connection differ by a large amount, the web of the column can be subjected to large shears. In such situations it may be necessary to provide shear stiffening in the form of diagonal stiffeners or double plates. Design of such stiffeners is treated in many design handbooks.[13.10, 15.16, 17.6]

18.5 DESIGN RECOMMENDATIONS

Depending on the anticipated behavior, bolted beam-to-column connections are designed either as slip-resistant or bearing-type joints. The design recommendations in Chapter 5 for fasteners in butt joints are also applicable to the design of bolted beam-to-column connections.

The bolts in an end-plate connection are subjected to combined tension and shear. The elliptical interaction curve for bolts subjected to combined loading conditions (see Eq. 4.8) can be used to examine the adequacy of the fasteners.

With the exception of end-plate connections, it can be assumed for design that the web connection or the seat angle transfers the shear component. Web shear connections should be designed as eccentrically loaded joints in accordance with the recommendations given in Chapter 13. The moment on a beam-to-column connection is transferred by structural components connected to the beam and column flanges. The recommendations given in Chapter 16 for beam and girder splices are applicable to the design of the beam-flange connection. The tension connection between the beam flange and column flange is usually critical for design. Prying forces should be considered for the design of the fasteners as well as joint components. The bolts and end plate adjacent to the tension flange can be treated as an equivalent tee stub connection, loaded in tension. Design recommedations for the T-stub connection are given in Chapter 17.

Special attention should be given to the bending stiffness of the column flanges to which the T-section or the end plate is fastened. The deformations of the column flanges and the T-section (end plate) may introduce prying forces (see Chapter 17) depending on their stiffness.

Stiffening the column may be required to prevent premature failure of a joint component due to column web crippling or column flange deformation. For connections with flange splice plates welded to the column the requirements for welded connections are applicable.[2.11],[18.8] If the compression flange force is transferred through an end plate or a T-section, Eq. 18.1 can be used to determine whether additional column stiffening is needed.

$$\sigma_{yc} w_c (Q + 5k_c) \geq A_f \sigma_{yb} \tag{18.1}$$

For slender webs, stability of the compression region may govern rather than strength alone. Reference 18.10 has suggested that the following relationship be satisfied when $d_c/w_c > 180 \sqrt{\sigma_{yc}}$

$$w_c^3 \leq \frac{\sigma_{yb}}{\sigma_{yc}} A_f \frac{d_c \sqrt{\sigma_{yc}}}{4100} \tag{18.9}$$

where d_c is the column web depth

The flanges of the column must not permit excessive deformations caused by the concentrated flange tensile forces. If splice plates are welded to the column, adequate resistance is provided by

$$t_{fc} \geq 0.4 \left(A_f \frac{\sigma_{yb}}{\sigma_{yc}} \right)^{1/2} \tag{18.2}$$

For bolted T-connections in tension (including end plate connections) the column flanges are adequate when

$$\left(\frac{T}{2}\right)\left(\frac{g}{2}\right) \leq \bar{M} \tag{18.3}$$

where $\bar{M}$ is the permissible moment on the effective length b of the column flange. The effective length b is defined as

$$b = r + \frac{3g}{2} \tag{18.4}$$

Equations 18.1 to 18.4 are applicable to both allowable stress design and load factor design methods. For allowable stress design, the permissible moment $\bar{M}$ is equal to

$$\bar{M} = \frac{b t_{fc}^2}{6} 0.75 \sigma_{yc} \tag{18.5}$$

For load factor design, the permissible moment is equal to the plastic moment of the effective column flange length,

18.5 Design Recommendations

$$\bar{M} = \frac{bt_{fc}^2}{4} \sigma_{yc} \tag{18.6}$$

If stiffeners are needed, they must carry the excess concentrated flange forces that the column web and flange are unable to carry.

For stiffeners opposite the beam compression flange, the required stiffener area can be determined from equilibrium. This yields

$$\sigma_{ys}A_{st} = \sigma_{yb}A_f - w_c(Q + 5k_c)\sigma_{yc} \tag{18.7}$$

If $C_1 = \sigma_{yb}/\sigma_{yc}$ and $C_2 = \sigma_{yc}/\sigma_{ys}$, Eq. 18.7 can be expressed as:

$$A_{st} = [C_1 A_f - w_c(Q + 5k_c)]C_2 \tag{18.7a}$$

If Eq. 18.9 governs the column web thickness, the stiffener area becomes

$$A_{st} = \left[C_1 A_f - \frac{4100\, w_c^3}{d_c \sqrt{\sigma_{yc}}} \right] \sigma_{yc} \tag{18.10}$$

A comparable requirement can be developed for stiffeners opposite the tension flange by considering the needed additional flange area to be resisted by stiffeners. Equilibrium yields

$$\sigma_{ys}A_{st} = \sigma_{yb}A_f - \sigma_{yb}A_f' \tag{18.8}$$

where $A_f\sigma_{yb}$ is the beam flange tension force and $A_f'\sigma_{yb}$ is the beam flange tension force that does not require stiffeners. This latter force can be estimated from Eq. 18.2 for the column flange thickness furnished. This yields

$$A_f' = \frac{\sigma_{yc}}{\sigma_{yb}} t_{fc}^2 \frac{100}{16} \cong 6 \frac{\sigma_{yc}}{\sigma_{yb}} t_{fc}^2$$

Substitution into Eq. 18.8 yields

$$\sigma_{ys}A_{st} = \sigma_{yb}A_f - 6\sigma_{yc}t_{fc}^2 \tag{18.8a}$$

Hence the required stiffener area opposite the beam tension flange becomes

$$A_{st} = [C_1 A_f - 6t_{fc}^2]C_2 \tag{18.8b}$$

The fastener shear stresses and the bearing stresses suggested in Chapter 5 were shown in Refs. 18.7, 18.10 and 18.11 to be fully applicable to beam-to-column connections.

References

18.1 J. C. Rathbun, "Elastic Properties of Riveted Connections," *Transactions, ASCE*, Vol. 101, 1936.

18.2 W. H. Munse, W. G. Bell and E. Chesson, Jr., "Behavior of Riveted and Bolted Beam-to-Column Connections," *Journal of the Structural Division, ASCE*, Vol. 85, ST3 March 1959.

18.3 R. A. Hechtman and B. G. Johnston, *Riveted Semi-Rigid Beam-to Column Building Connections*, AISC, Progress Report 1, 1947.

18.4 J. R. Bailey, "Strength and Rigidity of Bolted Beam-to-Column Connections," *Proceedings, Conference on Joints in Structures*, University of Sheffield, Sheffield, July 1970.

18.5 J. O. Surtees and A. P. Mann, "End Plate Connections in Plastically Designed Structures," *Proceedings Conference on Joints in Structures*, University of Sheffield, Sheffield, July 1970.

18.6 A. N. Sherbourne, "Bolted Beam-to-Column Connections," *The Structural Engineer*, London, 1961.

18.7 J. S. Huang and W. F. Chen, *Steel Beam-to-Column Moment Connections*, Meeting Reprint 1920, ASCE National Structural Engineering Meeting, April 1973.

18.8 J. D. Graham, A. N. Sherbourne, A. N. Khabbaz, and C. D. Jensen, *Welded Interior Beam-to-Column Connections*, AISC, New York, 1959.

18.9 C. W. Lewitt, E. Chesson, Jr., and W. H. Munse, *Restraint Characteristics of Flexible Riveted and Bolted Beam-to-Column Connections*, Engineering Experiment Station Bulletin 500, University of Illinois, 1969.

18.10 W. F. Chen and D. E. Newlin, "Column Web Strength in Steel Beam-to-Column Connections," *Journal of the Structural Division*, ASCE, Vol. 99, ST9, September 1973.

18.11 J. S. Huang, W. F. Chen and L. S. Beedle, "Behavior and Design of Steel Beam-to-Column Moment Connections, Welding Research Council Bulletin 188, New York, October 1973.

AUTHOR INDEX

Abolitz, A. L.
  (13.6)   211
Albrecht, P.
  (5.51)   117, 134
Allan, R. N.
  (4.26)   61, 75, 174, 175, 176, 179, 181
American Association of State Highway Officials
  (2.2)   16, 26
American Institute of Steel Construction
  (2.11)   26, 35, 126
  (13.7)   215
American Railway Engineering Association
  (2.1)   16, 26
AREA Committee on Iron and Steel Structures
  (4.14)   56
  (6.1)   143
American Society for Testing and Materials
  (1.3)   1, 2, 3, 4, 5, 37, 41
  (1.9)   1, 3, 5, 37, 41
  (1.10)   3, 37
  (1.11)   5
  (1.12)   7
  (2.6)   23, 116
  (2.9)   25
  (2.14)   23
ASCE – Manual 41
  (17.6)   268, 269, 299
ASCE – Manual 48
  (1.6)   1
American Welding Society
  (12.11)   199, 200
Aroian, L. A.
  (5.33)   129
Bailey, J. R.
  (18.4)   281, 292, 295
Ball, E. F.
  (4.16)   56
Baron, F.
  (3.6)   30, 118
  (3.7)   30, 118
Barsom, J. M.
  (2.20)   25
Bateman, E. H.
  (1.1)   1
Batho, C.
  (1.1)   1
Beano, S. Y.
  (12.7)   194
Beedle, L. S.
  (4.7)   32, 49, 52, 58, 75, 84, 86, 188
  (5.49)   123
Bell, W. G.
  (18.2)   281, 286
Bendigo, R. A.
  (4.6)   32, 49, 52, 58, 75, 84, 86, 88, 94, 172
  (5.35)   83

Bendigo, R. A. (cont'd)
  (8.1)   172
Benjamin, J. R.
  (2.13)   17
Birkemoe, P. C.
  (4.11)   64, 194, 195, 207, 208
  (5.20)   116, 118, 120
  (5.42)   118, 120
  (12.13)   195, 196, 197, 207, 208
  (15.3)   243, 244
  (17.4)   257, 261, 266, 267, 268, 270
Bouwman, L. P.
  (5.4)   75
  (5.36)   83
Boyd, W. K.
  (4.23)   65, 66
Brady, W. G.
  (5.27)   105
Brookhart, G. C.
  (4.18)   63, 64, 194, 195, 196, 199
  (5.9)   75

Cannon, J. C.
  (16.3)   250, 252, 254, 281
Carter, J. W.
  (5.48)   119
Centre de Recherches Scientifiques et Techniques de l'Industrie des Fabrications Metalliques (CRIF), Section Construction Metallique
  (12.5)   198, 199
Chang, W. N.
  (11.2)   190, 191, 192, 193
Chen, C. C.
  (5.14)   75
  (10.2)   187
Chen, W. F.
  (14.4)   233, 281
  (18.7)   281, 301
Chesson, E., Jr.
  (4.3)   44, 46
  (4.8)   52, 53
  (4.9)   53, 58, 60, 61, 65, 116, 118
  (4.20)   67
  (4.21)   67
  (5.7)   75
  (5.13)   75
  (5.28)   143, 145, 146, 147, 150, 152
  (5.29)   143, 145, 146, 147, 150, 152
  (5.45)   84

  (6.3)   143, 145, 146
  (17.2)   257, 261, 263, 265, 275
  (18.2)   281, 286, 287
  (18.9)   287
Chiang, K. C.
  (5.15)   75, 77
  (5.34)   75
Chin, A. G.
  (5.3)   75
Christopher, R. J.
  (4.1)   44, 59, 62, 81
Cochrane, V. H.
  (5.26)   105
Cornell, C. A.
  (2.13)   17
Cox, H. C.
  (3.2)   30, 31, 32, 33
Crawford, S. F.
  (13.2)   211, 213, 215, 218, 219, 220, 222, 225
Croth, N. S.
  (4.29)   60

Davis, C. S.
  (15.7)   241, 245
Davis, E.
  (7.3)   158, 160
Davis, H. E.
  (7.3)   158, 160
de Back, J.
  (5.4)   75
  (5.36)   83
  (5.39)   109
  (17.3)   257, 261, 263, 268, 270, 275
  (17.5)   264
De Jonge, A. E. R.
  (1.8)   1
de Jong, A.
  (5.39)   109
Desai, S.
  (7.2)   158, 159, 164
Deutscher Stahlbau-Verband
  (1.7)   2
Dineen, R. L.
  (4.20)   67
Divine, J. R.
  (5.7)   75
  (5.13)   75
Douty, R. T.
  (16.2)   250, 251, 252, 253, 257, 261,

Douty, R. T. (cont'd)
    268, 269, 275, 281, 290, 291, 292
Drew, F. P.
  (4.15)   56
Drucker, D. C.
  (5.27)   105

Elsea, A. R.
  (4.23)   65, 66
Eubanks, R. A.
  (15.3)   243, 244
European Convention for Constructional Steelwork
  (5.30)   73, 194, 298
  (9.1)   174

Faustino, N. L.
  (4.8)   52, 53
Fernlund, I.
  (5.47)   114
Fisher, J. W.
  (2.21)   22
  (3.8)   33, 158, 165, 192
  (4.1)   44, 59, 62, 81
  (4.2)   44, 58, 62, 73
  (4.3)   44, 59
  (4.4)   48, 84, 218, 219, 220
  (4.7)   32, 49, 52, 58, 75
  (4.25)   52
  (4.26)   61, 75, 174, 175, 176, 179, 181
  (4.28)   60
  (5.6)   76, 84, 86, 94, 97
  (5.10)   75, 184, 185, 186
  (5.11)   75, 194, 195, 201
  (5.12)   77, 84, 86, 94
  (5.21)   77, 86, 90, 97
  (5.22)   93, 159, 218, 219
  (5.24)   97
  (5.25)   84, 86, 97
  (5.49)   123
  (5.51)   117, 134
  (7.1)   154, 155, 159, 160, 161, 165
  (7.2)   158, 159, 164
  (7.5)   155, 156, 163
  (8.1)   172
Flint, T. R.
  (5.2)   75
Foreman, R. T.
  (4.5)   32, 49, 58, 75

Frank, K. H.
  (2.21)   22
Freudenthal, A. M.
  (2.17)   17
Frincke, M. H.
  (4.17)   56

Galambos, T. V.
  (2.23)   18
Gaylord, C. N.
  (15.6)   242, 243, 281
Gaylord, E. H.
  (15.6)   242, 243, 281
Gibson, G. J.
  (6.5)   145, 151
Graham, J. D.
  (18.8)   295, 297, 298, 300
Gurney, T. R.
  (5.43)   120
  (2.24)   22

Haisch, W. T.
  (8.2)   172, 174
Hansen, N. G.
  (5.19)   118
Hansen, R. M.
  (4.6)   32, 52, 58, 75, 84, 86, 88, 94, 172
  (5.35)   83
Haugen, E. B.
  (2.4)   17
Hechtman, R. A.
  (3.3)   30, 31
  (5.2)   75
  (5.3)   75
  (18.3)   281, 287
Herrschaft, D. C.
  (4.11)   64, 194, 195, 198, 207, 208
Higgins, J. J.
  (4.16)   56
Higgins, T. R.
  (3.5)   32, 33
  (13.4)   211, 212, 213, 215
  (13.9)   215
Hirano, M.
  (5.40)   109, 112
Hirt, M. A.
  (2.21)   22
Hojarczyk, S.
  (5.37)   194, 199

Holz, N. M.
(14.2) 233, 235, 236
Hoyer, W.
(14.1) 233, 235
Howard, L. L.
(4.13) 55, 83
Huang, J. S.
(14.4) 233, 281
(18.7) 281
Humphrey, K. D.
(4.29) 60
Hyler, W. S.
(4.23) 65, 66
(4.29) 60

Industrial Fasteners Institute
(1.5) 5
International Association for Bridge and Structural Engineering
(2.18) 17
Irwin, G. P.
(2.10) 23
Ives, K. D
(8.3) 172, 174

Jensen, J. D.
(18.8) 295, 297, 298, 300
Jones, J.
(5.32) 109, 111
Johnson, L. G.
(10.1) 184, 185
(16.3) 250, 252, 254, 281
Johnson, K. L.
(5.41) 114
Johnston, B. G.
(3.1) 32
(18.3) 281, 287

Kasinski, J.
(5.37) 194, 199
Kennedy, D. J. L.
(12.10) 207, 208
Khabbaz, A. N.
(18.8) 295, 297, 298, 300
Klingerman, D. J.
(5.51) 117, 121, 134
Klöppel, K.
(5.18) 113, 116, 118, 120
(6.2) 143, 151, 172
Koepsell, P. L.

(5.2) 75
Kormanik, R.
(5.24) 97
Kraft, J. M.
(2.10) 23
Krickenberger, C. F., Jr.
(4.21) 67
Kulak, G. L.
(4.1) 44, 59, 62, 81
(5.12) 75, 77, 84, 86, 94
(5.23) 97, 99, 100
(5.25) 84, 86, 97
(5.38) 100
(13.2) 211, 213, 215, 218, 219, 220
(13.12) 220, 228
(14.2) 233, 235, 236
Kuperus, A.
(5.8) 75

Larson, E. W., Jr.
(3.6) 30, 118
(3.7) 30, 118
Laub, W. H.
(5.1) 75
Lee, J. H.
(5.10) 75, 184, 185, 186
(5.11) 75, 77, 194, 195, 201
Lenzen, K. H.
(5.48) 114
Lewitt, C. W.
(18.9) 287
Lind, N. C.
(2.19) 17
Lindner, C. W.
(6.4) 143, 145, 150
Lobb, V.
(12.9) 205, 206
Lu, Z. A.
(5.16) 75

Macadam, J. N.
(4.22) 66
Maney, G. A.
(4.12) 55
Mann, A. P.
(18.5) 281, 292, 293, 294
Maseide, M.
(5.17) 75, 77, 194, 198, 199, 207
McCammon, L. B.
(6.4) 143, 145, 150

Author Index

McGuire, W.
(13.11) 219
(16.2) 250, 251, 252, 253, 257, 261, 268, 269, 275, 281, 290, 291, 292
McNamee, B. M.
(2.21) 22
(5.51) 117, 121, 134
Meinheit, D. F.
(5.20) 114, 118, 120
Möehler, K.
(5.5) 75, 77, 84, 118, 233, 235
(9.2) 174, 194, 203, 204, 206, 207, 233, 235, 236, 238, 239
Munse, W. H.
(3.2) 30, 31, 23, 33
(3.5) 32, 33
(4.8) 52, 53
(4.9) 53, 58, 60, 61, 65, 116, 118
(4.19) 63
(4.20) 67
(4.21) 67
(4.27) 64, 194, 198, 207
(5.7) 75
(5.13) 75
(5.20) 114, 118, 120
(5.28) 143, 145, 146, 147, 150, 152
(5.29) 143, 145, 146, 147, 150, 152
(5.31) 109, 111, 112
(6.3) 143, 145, 146
(12.13) 195, 196, 197, 207, 208
(15.3) 243, 244
(17.1) 243, 244, 257, 266
(17.2) 257, 261, 263, 265, 275
(17.4) 257, 261, 266, 267, 268, 270
(18.2) 281, 286
(18.9) 287

Nadai, A.
(5.46) 77, 102
Nair, R. S.
(17.4) 257, 261, 266, 267, 268, 270
Nawrot, T.
(5.37) 194, 199

O'Connor, C.
(5.11) 75, 77, 194, 195, 201
O'Connor, J. J.
(5.41) 114
Office of Research and Experiments of the
International Union of Railways (ORE)
(12.1) 194, 195, 196, 199, 203
(12.2) 194, 200
(12.3) 194, 205, 206
(12.4) 207
Oliver, W. A.
(3.4) 32
Oyeledun, A. O.
(4.28) 60

Paris, P. C.
(2.10) 23
(2.15) 22
Pauw, A.
(4.13) 55, 83
Pellini, W. S.
(2.8) 25
Peterson, K. S.
(17.2) 257, 261, 263, 265, 275
Phillips, J. R.
(5.1) 75
Power, E.
(7.1) 154, 155, 156, 159, 160, 161, 165

Ramseier, P.
(4.7) 32, 49, 52, 58, 75, 84, 86, 88
Rathbun, J. C.
(18.1) 281, 282, 284, 286, 288
Reemsnyder, H. S.
(2.7) 23
Regec, J. E.
(14.4) 233
Reilly, C.
(13.1) 211
Research Council on Riveted and Bolted Structural Joints of the Engineering Foundation
(1.4) 2, 3
(2.12) 26
Rivera, U.
(7.5) 155, 156, 163
Rolfe, S. T.
(2.20) 25
Rumpf, J. L.
(4.2) 44, 45, 46, 58, 62, 73, 74
(4.5) 32, 49, 58, 75
(4.6) 32, 52, 58, 75, 84, 86, 88, 94
(5.21) 77, 86, 93, 97
(5.35) 83

Rumpf, J. L. (cont'd)
  (8.1)   172
  (17.7)   257
Rust, T. H.
  (15.1)   241, 243, 244
Ruzek, J. M.
  (2.16)   26

Salmon, C. G.
  (3.1)   32
Savikko, E. R.
  (5.3)   75
Schaaf, T. v. d.
  (12.6)   199, 203, 205, 206, 207
Schenker, L.
  (3.1)   32
Schutz, F. W., Jr.
  (16.1)   250, 252, 254, 281, 290, 292
Scott, M. B.
  (6.4)   143, 145, 150
Seeger, T.
  (5.18)   113, 116, 120
  (6.2)   143, 151, 172
Selberg, A.
  (5.17)   75, 77, 194, 198, 199, 207
Sherbourne, A. N.
  (18.6)   281, 292, 295
  (18.8)   295, 297, 298, 300
Shermer, C. L.
  (13.5)   211
  (13.8)   228
Shourkry, Z.
  (8.2)   172, 174
Siddiqi, I. H.
  (4.18)   63, 194, 195, 196, 199
  (5.9)   75
Skwirblies, H.
  (14.1)   233, 235
Slutter, R. G.
  (13.3)   211, 213
Spooner, L. A.
  (16.3)   250, 252, 254, 281
Srinivasan, R. S.
  (5.42)   118, 120
Staff Fritz Engineering Laboratory
  (14.4)   233
Steel Structures Painting Council (SSPC)
  (12.12)   201, 202, 204
Steinhardt, O.
  (5.5)   75, 77, 84, 118, 233, 235
  (9.2)   174, 194, 203, 204, 206, 207, 233, 235, 236, 238, 239
Sterling, G. H.
  (4.3)   44, 49
  (5.6)   75, 76, 84, 86, 94, 97
Stoller, F.
  (12.9)   205, 206
Stout, R. D.
  (2.16)   26
Strating, J.
  (2.22)   18
Structural Engineers Association of California
  (2.3)   17
Struik, J. H. A.
  (4.28)   60
  (15.8)   241, 245, 246
  (17.3)   257, 261, 263, 268, 270, 275
Subcommittee on Bolt Strength
  (4.24)   65, 66
Surtees, J. O.
  (18.5)   281, 292, 293, 294

Tajima, J.
  (4.10)   61
  (5.44)   122
Tall, L.
  (13.10)   214, 281, 288, 299
Thomas, F. P.
  (1.2)   2
Tomonaga, K.
  (5.44)   122
Tör, S. S.
  (2.16)   26
Troup, E. W. J.
  (4.3)   44, 46

Valtinat, G.
  (9.2)   174, 194, 203, 204, 206, 207, 233, 235, 236, 238, 239
van Douwen, A. A.
  (5.4)   75
Vasishth, U. C.
  (5.16)   75
Vasarhelyi, D. D.
  (4.18)   63, 194, 195, 196, 199
  (5.9)   75
  (5.14)   75
  (5.15)   75, 77
  (5.16)   75
  (5.34)   75

Author Index

Vasarhelyi, D. D. (cont'd)
  (10.2)   187
  (11.1)   190, 191, 193
  (11.2)   190, 191, 192, 193
  (12.7)   194
  (15.4)   241, 243, 244, 245
Vincent, G. S.
  (5.50)   123, 128
Viner, J. G.
  (4.20)   67

Waddell, J. A. L.
  (15.5)   241, 242, 243
Wake, B. T.
  (6.5)   145, 151
Wallaert, J. J.
  (4.4)   48, 84, 218, 219, 220
  (4.25)   52
Weibull, W.
  (2.5)   23, 116
Wells, A. A.
  (2.10)   23
Whitmore, R. E.
  (15.2)   241, 243, 244, 246
Wilson, W. M.
  (1.2)   2
  (3.4)   32
Woodruff, G. B.
  (7.3)   158, 160
Wyly, L. T.
  (5.48)   114
  (6.4)   143, 145, 150

Yarimci, E.
  (13.3)   211, 213
Yen, B. T.
  (5.51)   117, 121, 134
Yoshida, N.
  (3.8)   33, 158, 165, 192
Young, D. R.
  (5.3)   75
Ypeij, E.
  (14.3)   236, 239
Yusavage, W. J. (ed.)
  (7.4)   164

Zennaro, L.
  (12.8)   197, 198
Zoetemeyer, P.
  (17.5)   264

SUBJECT INDEX

AASHO Specification, 16, 152
Allowable Stresses
  Concepts, 26
  Bolted Butt Joints, 123, 127, 132, 133, 135, 137
Alignment of Holes, 130
American Society for Testing and Materials
  ASTM A325, 1, 3, 4, 5
  ASTM A490, 1, 3, 4, 5
  ASTM A307, 3, 4, 5
  ASTM A502, 7
AREA Specification, 16
Axial Shear, 11, 169

Beam Splices, 250
Beam-to-Column Connections, 281
  Types, 282
  Load Deformation Behavior, 285-294
  Stiffening Requirements, 294-299
  Design Recommendations, 299
Bearing, 84, 85, 108, 109, 111, 112
Bearing Ratio, 111, 112
Bearing Stresses, 109, 135
Bearing Type Joint
  Definition, 20
  Load Deformation Behavior, 84
  Load Partition, 90, 158
  Stiffness, 88
  Surface Preparation, 89, 194
Bolted Joints
  General Behavior, 19, 84, 145, 154, 169, 179
  Stiffness, 75, 88
Bolt Relaxation, *see* Relaxation
Building Codes, 16
Built-Up Sections
  General, 143
  Load Deformation Behavior, 145, 146, 148, 150
  Shear Lag, 145, 147, 149, 150
  Design Recommendations, 151, 152
Butt Splice
  Symmetric, 15, 72
  Load Deformation Behavior, 72, 84
  Unbuttoning, 86
  Design Recommendation, 123, 133

Calibrated Wrench Method, 55, 83, 131
Clamping Force, *see also* Tension Type Connections
  Bolts, 55
  Calibrated Wrench Method, 83
  Effect of Hole Dimensions, 174, 175
  Relaxation, 61, 179
  Rivets, 30
  Slip Resistance, 72, 80
  Turn-of-Nut Method, 55, 62, 79, 80
  Tension Indicators, 60, 74
Coated Surfaces, *see* Surface Treatment; Surface Preparation

Subject Index

Connections
  Types, 11
  Flat Plate Type, 15
  Symmetric Butt Splice, 15, 72, 90
  Gusset Plate, 15, 241
  Lap Plate Splice, 15, 169
  Bracket Type, 15, 211
  Hanger Type, 15, 265
  Shingle Joints, 154
Combination Joints
  General, 231
  Load Deformation Behavior, 233, 234, 236, 237
  Design Recommendations, 239
Corrosion of Bolts, 66

Design Recommendations
  Beam - Girder Splices, 256
  Beam-to-Column Connections, 299
  Built-Up Sections, 151, 152
  Butt Joints, 122-138
  Coated Joints, 208
  Combination Joints, 239
  Filler Plates, 188
  Gusset Plates, 246
  Lap Joints, 173
  Oversize Holes, 181
  Riveted - Bolted Joint, 239
  Shingle Joints, 163, 165, 167
  Single Bolts, 68, 69
  Single Rivets, 35
  Tension Type Connections, 275
  Welded–Bolted Joint, 231, 233
Direct Tension Indicators, 55, 59, 60, 74

End Distance, 108, 111, 136, 137
Eccentric Shear, 15, 147, 155
Eccentric Connections, *see* Built-Up Sections; Lap Joints; Gusset Plates; and Beam-to-Column Connections
Eccentrically Loaded Joints
  General, 211
  Load Deformation Behavior, 211, 212, 213
  Analysis, 214, 215-219
  Design Recommendations, 220-230

Factor of Safety, 17, 123
Failure Probability, 17, 18, 116
Fatigue
  Bolts in Tension, 266
  Bearing Type Joints, 120, 208
  Butt Joints, 116, 206, 239, 240
  Coated Joints, 207
  Combination Joints, 239
  Design Recommendations, 134, 208
  Failure Modes, 112
  Failure Probability, 16, 116
  Fracture Mechanics, 23, 24
  Fretting Fatigue, 116, 120
  S - N Curves, 22, 23, 117, 118
  Slip Resistant Joints, 118, 206, 207
Filler Plates
  Types, 184, 185
  Design Recommendations, 188
Fracture, 23
Fretting Fatigue, 116, 120

Galvanized Bolts, 63, 194
Galvanized Joints, 194, 195, 205, 207
Girder Splices, 250
Grip Length, 32, 46, 59, 83, 107
Gusset Plates
  General, 15, 143, 241
  Analysis, 242, 243, 244, 245

Hanger Connection, 15, 257
High Strength Bolts
  Galvanized, 63, 194
  History of, 1
  Identification, 38, 39
  Load Deformation Characteristics
    Direct Tension, 42-46
    Torqued Tension, 42-46
    Shear, 47-52
  Mechanical Properties, 3, 4, 5
  Relaxation, 61, 179
  Reuse, 62
  Specifications, 2, 231
  Strength - Tension, 41, 68
    Shear, 47, 68
    Combined Tension - Shear, 53, 69
  Types, 3, 37

Installation Procedure
  Effect of Hole Dimensions, 175
  High Strength Bolts, 55, 56, 57
  Oversize Holes, 174
  Rivets, 29
  Slotted Holes, 174

Subject Index

Interference Body Bolts, 37, 40

Joint, *see also* Connections
Joint Efficiency
  Built-Up Sections, 145
  Butt Joints, 94
Joint Stiffness, 75, 88, 158

Lap Joints
  General, 143
  Load Deformation Behavior, 173
  Design Recommendations, 173
Length of Joint
  Influence on Load Partition, 95, 146, 159
Load Factor Design, 17, 18
  Beam-to-Column Connections, 300
  Butt Joints, 124
  Eccentrically Loaded Joints, 227
  Tension Type Connections, 277
Load Partition
  Butt Joints, 72, 90
  Shingle Joints, 158
Load Transfer
  Butt Joints, 71, 84, 145
  Shear and Bearing, 20
  Friction, 19, 71, 205
  Shingle Joints, 154
Load Types, 16

Mechanical Properties
  Bolts, 3
  Rivets, 3, 7
Metallized Joints, 194, 195, 196, 205, 207
Misalignment of Holes, 190, 191, 192

Net Section
  Strength, 99, 104, 145, 146, 147, 148, 245, 254
  Design Recommendations, 133, 152, 167, 246, 256
Nuts
  Specifications, 5, 37
  Strength, 67
  Use of Nuts, 67

Oversize Holes
  Effect on Installation, 174
  Dimensions, 174

Painted Surfaces, 77, 78, 194, 201
Prying Action, *see* Tension Type Connection

RCRBSJ Specification, 2, 3, 55, 58, 67, 174, 195
Relaxation of Clamping Force, 61, 178
Rivets
  Clamping Force, 30, 31
  Design Recommendations, 35
  Installation, 29
  Mechanical Properties, 7, 29, 31
  Specifications, 7
  Strength, 31, 32, 34
Riveted - Bolted Joints, 231, 238

Shear
  Axial, 11, 15
  Eccentric, 11, 15
  Shear Resistance of Bolts, 48
Shear Lag in Built-Up Sections, 145, 147, 149, 150
Shingle Joints
  General, 154
  Load Deformation Behavior, 154, 155, 156
  Stiffness, 158
  Load Partition, 158
  Design Recommendations, 163
Slip Coefficient, 75
Slip Test, 72
Slip of a Connection, 19, 71
Slip Resistant Joints, 19, 20, 71
Slip Resistance
  Basic, 71, 72, 73, 74, 75-78, 172, 179
  Effect of Clamping Force, 79
  Effect of Joint Geometry, 74, 185
  Effect of Number of Faying Surfaces, 72, 154
  Effect of Hole Dimensions, 175, 190
Stiffness, *see* Joint Stiffness
Stress Area, 5
Structural Steel
  Types, 9
  Mechanical Properties, 9, 10, 11
Stress Corrosion Cracking, 64, 66
Surface Preparation, 75, 76, 194
Surface Treatment
  Load Deformation Behavior, 195, 205, 206

Swedge Bolts, 37

Tension Shear Jig, 48, 73
Tension Type Connection
  Fastener in Tension, 257-260
  Fastener Group in Tension, 260
  Repeated Loading, 266
  Prying Action, 268-274
  Design Recommendations, 275-279
Torque - Tension Relationship, 42, 43, 44
Turn-of-Nut Method, 55, 56, 80

Ultimate Strength of Joints
  Effect of Joint Geometry, 90, 148, 158
  Effect of Hole Dimensions, 175
  Ultimate Strength of Fastener Groups, 215, 218
  Unbuttoning, 86

Washers
  Load Indicator, 60
  Use of Washers, 65

Zinc Paints
  Slip Resistance, 195, 205, 206